Astronomy: A Study of Celestial Objects

Astronomy: A Study of Celestial Objects

Daniel Harris

MURPHY & MOORE
www.murphy-moorepublishing.com

Murphy & Moore Publishing,
1 Rockefeller Plaza,
New York City, NY 10020, USA

Visit us on the World Wide Web at:
www.murphy-moorepublishing.com

ISBN: 978-1-63987-061-5 (Hardback)

Cataloging-in-Publication Data

Astronomy : a study of celestial objects / Daniel Harris.
p. cm.
Includes bibliographical references and index.
ISBN 978-1-63987-061-5
1. Astronomy. 2. Space sciences. 3. Celestial sphere. I. Harris, Daniel.
QB47 .A88 2022
520--dc23

TABLE OF CONTENTS

PREFACE

It is with great pleasure that I present this book. It has been carefully written after numerous discussions with my peers and other practitioners of the field. I would like to take this opportunity to thank my family and friends who have been extremely supporting at every step in my life.

The domain of natural science that studies celestial objects and phenomena is known as astronomy. This discipline aims to study the evolution and origin of astronomical object such as planets, moons, galaxies, comets, stars, nebulae, etc. The branch of astronomy concerned with the recording of data of the observable universe using physical models is known as observational astronomy. Theoretical astronomy describes astronomical objects and astronomical phenomena using analytical models of physics and chemistry. There are various sub-fields in astronomy, such as astrophysics, astrobiology, physical cosmology, astrochemistry, solar astronomy, stellar astronomy, etc. This book attempts to understand the multiple branches that fall under the discipline of astronomy and how such concepts have practical applications. The topics included herein are of utmost significance and bound to provide incredible insights to readers. This book aims to serve as a resource guide for students and experts alike and contribute to the growth of the discipline.

The chapters below are organized to facilitate a comprehensive understanding of the subject:

Chapter - Introduction

Astronomy is the branch of science that deals with the study of celestial objects and its phenomena. It studies the origin and evolution of planets, moons, stars, nebulae, galaxies, and comets via the principles of mathematics, physics and chemistry. This is an introductory chapter which will briefly introduce about astronomy.

Chapter - Sub-disciplines of Astronomy

There are many sub-disciplines that fall under astronomy such as radio astronomy, exoplanetology, astrobiology, infrared astronomy, visible-light astronomy, ultraviolet astronomy, gamma-ray astronomy, etc. This chapter has been carefully written to provide an easy understanding of the various sub-disciplines of astronomy.

Chapter - Solar Astronomy: Study of the Sun

Solar astronomy is the branch of astronomy that is concerned with the study of sun and its phenomena. Solar physics, solar irradiance, standard solar model, etc. are some of the aspects that come within it. This chapter closely examines solar astronomy to provide an easy understanding of solar astronomy.

Chapter - Cosmology: A Comprehensive Study

Cosmology is the branch of astronomy which deals with the study of origin and evolution of universe. It is divided into physical cosmology, philosophical cosmology and religious cosmology. The Big Bang theory, dark matter and energy, are some of the concepts of cosmology. This chapter discusses all the concepts of cosmology in detail.

Chapter - Major Astronomical Tools

Major tools that are used in astronomy are astronomical telescopes, radio telescopes, astronomical clock, visible-light detector, etc. The topics elaborated in this chapter will help in gaining a better perspective of these astronomical tools.

Daniel Harris

Introduction

CHAPTER 1

Astronomy is the branch of science that deals with the study of celestial objects and its phenomena. It studies the origin and evolution of planets, moons, stars, nebulae, galaxies, and comets via the principles of mathematics, physics and chemistry. This is an introductory chapter which will briefly introduce about astronomy.

ASTRONOMY

Astronomy is a multidisciplinary branch of science that focuses on the study of the universe, celestial objects, and space. "Astronomy" is derived from Greek origin, with "astr" translating to star and "nomos" meaning arrangement. Astronomers have been able to hypothesize the answers to philosophical questions presented throughout history. Some of these questions include, "how did life begin," "is there life on other planets," and "did the big bang theory occur". The world today would not operate the same without the knowledge amassed throughout the centuries. For instance, cosmic events and weather patterns would be nothing but a mystery. To many, the idea of studying astronomy has very few practical implications. In reality, the astounding discoveries made by astrologists are being applied to improve our current technology and modern medicine.

Branches of Astronomy

Astronomy, defined as a multidisciplinary subject is composed of 17 sub branches.

Biology, physics, chemistry, mathematics, geology, and computer science are all fundamental aspects that intertwine with one another. Astrophysics and astrobiology are branches of astronomy that have resulted in significant technological advancements. By understanding the cosmic events that occur throughout the universe, humans are now aware of the impacts they have on our planet.

Astrobiology: Astrobiology focuses on the distribution, origin, and evolution of life in the universe. There are several basic but equally important concepts that astrobiologists are trying to understand. These ideas revolve around the origin of the building blocks of life and the role these elements play. Another important question to keep in mind is how do these compounds interact and combine with one another? The goal of astrobiology is to find different resources, minerals, and whether life exists on other planets. By applying this knowledge, there is potential to reconfigure many initial assumptions about the origin of life. Eventually, the creation of suitable living environments on Mars and other galaxies may be a possibility in the future.

Astrophysics: Astrophysics is a branch of astronomy that focuses on the physical properties and behavior of celestial bodies. Scientists can better understand their internal processes by applying the laws of chemistry and physics. Astrologists can utilize this information and determine if these celestial objects will have any negative impact on the earth. The patterns and characteristics of cosmic events also allow scientists to improve on current technology. These improvements can range anywhere from increasing the durability of satellites to sustain less damage from solar flares to utilizing modeling techniques developed in astrophysics to help diagnose dementia.

Astronomical Instruments

Tools used by astronomers to collect and record data fall into two distinct categories, observational and auxiliary equipment. Both types work in synch with one another and are required to analyze information accurately.

Observational Equipment: Observational equipment has two primary functions, determining exact coordinates and collecting radiation emitted by celestial bodies. Telescopes, photometers, spectrometers, and radio dishes are commonly used. Telescopes collect light and other forms of electromagnetic radiation. Filters, spectrometers, and photometers are special attachments that are used to resolve images and make them visible to the human eye.

Auxiliary Equipment: Auxiliary equipment records and analyzes all information collected from observational instruments. Determining the brightness of stars or accurately recording their spectra is just a few of their primary functions. Some examples are radiation/infrared detectors, diffraction gratings, and radio receivers.

Subfields of Astronomy

There are so many types of objects that astronomers study, that it is convenient to break astronomy up into subfields of study.

- One area is called planetary astronomy, and researchers in this subfield focus their studies on planets, both within and outside our solar system, as well as objects like asteroids and comets.
- Solar astronomy is the study of the Sun. The scientists who are interested in learning how it changes, and to understand how these changes affect the Earth, are called solar physicists. They use both ground-based and space-based instruments to make nonstop studies of our star.
- Stellar astronomy is the study of stars, including their creation, evolution, and deaths. Astronomers observe these objects across all wavelengths and apply the information to create physical models of the stars.
- Galactic astronomy focuses on the objects and processes at work in the Milky Way Galaxy. It's a very complex system of stars, nebulae, and dust. Astronomers study the motion and evolution of the Milky Way in order to learn how galaxies are formed.
- Beyond our galaxy lie countless others, and these are the focus of the discipline of extragalactic astronomy. Researchers study how galaxies move, form, break apart, merge, and change over time.
- Cosmology is the study of the origin, evolution, and structure of the universe in order to understand it. Cosmologists typically focus on the big picture and attempt to model what the universe would have looked like only moments after the Big Bang.

CELESTIAL OBJECTS

Our universe contains an amazing array of celestial objects, sometimes referred to as celestial bodies or astronomical objects. A celestial object is a naturally happening phenomenon that occurs in the observable universe. In astronomy, the word object and body are often used interchangeably. On the other hand, a celestial body is a solo, strongly bound, adjoining entity, while the celestial object is a complex, less cohesively bound structure, which may consist of multiple bodies or even other objects with substructures. Celestial bodies or heavenly groups are objects in space such as the sun, planets, moon, and stars.

They form a part of the massive universe we live in and are typically very far from us.

The magnicent night sky is dotted with such objects and when we see them using a telescope, they expose fascinating worlds of their own. Because they are so far away, we cannot see all of them with the naked eye and we depend upon telescopes to study them. The word celestial body is as wide as the entire universe, for both known and unknown. By denition, a celestial body is any natural object outside of the Earth's atmosphere. Simple examples are the Moon, Sun, and the other planets of our solar system. But those are very partial examples. The Kuiper belt holds many celestial bodies. Any asteroid in space is a celestial body.

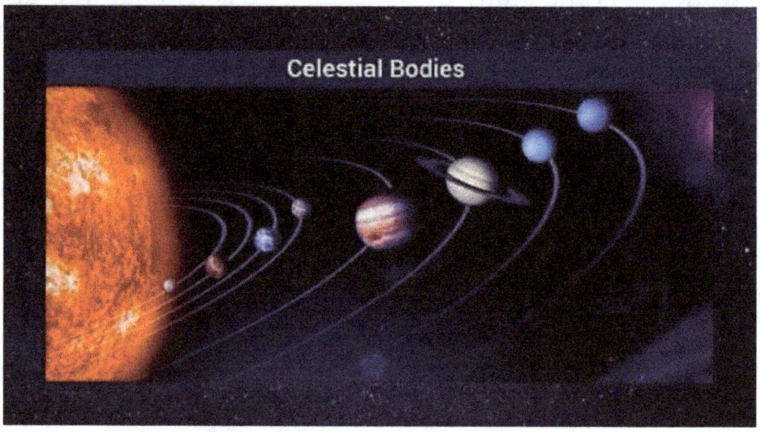

Classication of Celestial Bodies

Stars

A star is a form of a celestial object made up of a shining spheroid of plasma held together by its own gravity. The nearby star to Earth is the Sun. Several other stars are visible to the naked eye from Earth during the night time, looking at a multitude of fixed luminous points in the sky due to their enormous distance from Earth. Historically, the most noticeable stars were grouped into constellations and asterisms, the brightest of which gained proper tags. Astronomers have drawn together star catalogues that identify the known stars and deliver standardized stellar designations. However, it is estimated that there are over 300 sextillions (3×10^{23}) stars in the Universe, including all-stars outside our galaxy (Milky Way), are invisible to the naked eye from Earth.

A star's life stat with the gravitational collapse of a gaseous nebula of material composed mostly of hydrogen, along with helium and small amounts of heavier elements. When the lunar core is suciently thick, hydrogen becomes gradually converted into helium through nuclear fusion, liberating energy in the process. The rest of the interior of the star transfers energy away from the core through a mixture of the radiative and convective heat transfer process. The interior pressure prevents it from collapsing further under its own gravity. A star with mass bigger than 0.4 times the Sun's mass will expand to become a red hulk when the hydrogen fuel in its core is exhausted.

Planets

A planet is a body that revolves around a star that is enormous enough to be spherical by its own magnitude, is not huge enough to cause thermonuclear fusion, and has cleared its neighbouring region of planetesimals.

The word planet is an ancient word which ties to history, astrology, science, mythology, and religion. Five planets in the Solar System are seen with our naked eye. These were observed by many early cultures as celestial, or as emissaries of idols. As logical knowledge advanced, human awareness of the planets changed, incorporating several dissimilar objects.in our solar system, we have eight planets they are Mercury, Venus, Earth, Mars, Jupiter, Saturn, Uranus, Neptune.

Planets are mostly divided into two main types: big low-density giant planets, and smaller stony terrestrials. There are eight planets in the Solar System. In order of rising distance from the Sun, they are Mercury, Venus, Earth, and Mars, then the four giant planets, Jupiter, Saturn, Uranus, and Neptune. Six of the planets are circled by one or more natural satellites.

Numerous thousands of planets around other stars ("extrasolar planets" or "exoplanets") have been shown in the Milky Way. As of 5 Feb 2019, 3,956 known extrasolar planets in 2,973 planetary systems (plus 654 multiple planetary systems), going in size from just above the size of the Moon to gas goliaths about twice as large as Jupiter have been discovered, out of which more than 100 planets are of the same size as Earth, nine of which are at the same comparative distance from their star as Earth from the Sun, i.e. in the circumstellar habitable area.

Satellites

It is a natural celestial object with a recognized orbit around a planet of the Solar System, some as small as a kilometre across. In the Solar System, there are six terrestrial satellite systems covering 185 known natural satellites. Four IAU-Mentioned dwarf planets are also known to have natural satellites: Pluto, Haumea, Makemake, and Eris. As of September 2018, there are 334 other small planets known to have moons.

The Earth-Moon structure is unique in that the ratio of the mass of the Moon to the frame of Earth is much greater than that of any other natural-satellite–planet proportion in the Solar System (although there are minor-planet systems with even greater ratios, notably the Pluto–Charon system).

Comets

A comet is an icy object which is present in Solar System object that, when travelling close to the Sun, it warms and start to emit gases, a practice known as outgassing. This creates a visible atmosphere or coma, and sometimes also a tail. These occurrences are due to the eects of solar radiation and the solar wind acting upon the core of the comet. Comet core range from a few hundred meters to tens of kilometres across and are made up of loose collections of ice, dust, and small rocky particles. The coma can be up to 15 times the Earth's diameter, while the tail may give one astronomical unit. If satisfactorily bright, a comet may be seen from the Earth without the help of a telescope and may subtend an arc of 30° (60 Moons) through the sky. Comets have been seen and recorded since ancient times by many cultures.

Comets are distinguished from asteroidds by the existence of an extended, gravitationally unbound atmosphere nearby their central core. This atmosphere has parts named as the coma which is surrounded by its nuclei (the central part immediately surrounding the core) and the tail (a usually linear section consisting of dust or gas is blown emitting out from the coma by sun's RAYS pressure or out streaming solar air plasma). However, dead comets have passed d close to the Sun several times and have lost nearly all of their volatile ices and dust and may come to resemble minor asteroids. Asteroids are assumed to have a dierent origin than comets, having formed around Jupiter orbit rather than in the outer Solar System. The discovery of main-belt comets and lively centaur minor planets has a fuzzy distinction between asteroids and comets. In the rst period of the 21st century, there was the discovery of some minor bodies with longperiod comet orbits, but features of inner solar system asteroids were called Manx comets.

Asteroids

Asteroids are small planets, especially of the inside Solar System. Big asteroids are also called planetoids. These expressions have historically been applied to any astronomical body orbiting the Sun that did not look like a planet-like disc and was not seen to have characteristics of a lively comet such as a tail. As small planets in the outer Solar System were discovered they were naturally found to have volatile - rich tops like comets. As a result, they were frequently famed from objects found in the main asteroid belt, the word "asteroid" refers to the small planets of the inner Solar System.

There are zillions of asteroids, many assumed to be the crushed leftovers of planetesimals, bodies within the young Sun's solar nebula that never grew bigenough to bwcome planets. The massive majority of known asteroids orbit within the key asteroid belt located between the orbits of Mars and Jupiter is co-orbital with jupiter (the Jupiter Trojans). However, several other orbital families exist with signicant populations, including the near-Earth objects. Single asteroids are categorized by their typical spectra, with the majority falling into three key groups: C-type, Mtype, and S-type. These were termed after and are usually identied with carbon-rich, metallic, and silicate (stony) congurations, respectively. The sizes of asteroids vary greatly; the largest, Ceres, is almost 1,000 km (625 mi) across.

Asteroids are separated from comets and meteoroids. In the case of comets, the dierence is one of composition: while asteroids are mainly made of mineral and rock, comets are mainly composed of dust and ice. Furthermore, asteroids are formed closer to the sun, preventing the progress of cometary ice. The dierence between asteroids and meteoroids is mainly of size: meteoroids have a radius of one meter or less, whereas asteroids have a radius greater than one meter. Finally, meteoroids can be made of either cometary or asteroidal materials.

Meteor and Meteorites

A meteor, also known as a shooting star, the path of meteor is visible and glowing meteoroids glowing meteoroid, comet or asteroid through Earth's atmosphere, after being heated to burning by collisions with air molecules in the upper atmosphere, making a streak of light via it's quick motion and sometimes also by flaking growing material in it. Although a meteor may seem like few thousand feet from the earth meteor naturally occur in the mesosphere at altitudes from 76 to 100 km (250,000 to 330,000 ft).

Billions of meteors enter Earth's atmosphere daily. Most meteoroids that cause meteors are about the size of a particle of sand, i.e. they are usually millimeter-sized or even smaller. Meteoroid sizes can be measured from their mass and density which, in turn, can be expected from the observed meteor trajectory in the higher atmosphere Meteors showers is a natural phenomenon and it can occur in showers, which begins when Earth travell through a tributary of debris left by a comet, or as "random" or meteors, not associated with a specic stream of space debris. Several specic meteors have been seen, largely by members of the public and other largely by accident, but with enough information that orbits of the meteoroids producing the meteors have been measured. The atmospheric speeds of meteors result from the movement of Earth around the Sun at about 30 km/s (68,000 mph), the orbital speeds of meteoroids, and the gravity well of Earth.

Galaxies

A galaxy is a gravitational system of stars, interstellar gas, stellar fragments, dust, and dark matter. Galaxies are in size from small with just a few hundred million (10^8) stars to colossi with one hundred trillion (10^{14}) stars, each orbiting its galaxy's center of mass.

Galaxies are characterized according to their visual morphology as oval, spiral, or irregular. Many galaxies are believed to have supermassive black holes at their cores. The Milky Way's central black hole, known as Sagittarius A*, has a weight of four million times greater than the Sun. Since April 2016, GNz11 is the oldest and best reserved observed galaxy with a comoving distance of 32 billion light-years from Earth and observed as it existed just 400 million years after the Big Bang.

The space between galaxies is lled with an unsubstantiated gas (the intergalactic medium) having an average mass of less than one atom per cubic meter. Most galaxies are gravitationally systematized into groups, clusters, and superclusters. The Milky Way is part of the Local Group, which is ruled by it and the Andromeda Galaxy and is part of the Virgo Supercluster. At the biggest scale, these associations are mostly arranged into sheets and laments surrounded by immense spaces. The biggest structure of galaxies yet to recognize as a cluster of superclusters that has been termed Laniakea, which holds the Virgo supercluster.

SUPERNOVA

A supernova is the explosion of a star. It is the largest explosion that takes place in space.

A supernova is a large explosion that takes place at the end of a star's life cycle.

Supernovas are often seen in other galaxies. But supernovas are difficult to see in our own Milky Way galaxy because dust blocks our view. In 1604, Johannes Kepler discovered the last observed supernova in the Milky Way. NASA's Chandra telescope discovered the remains of a more recent supernova. It exploded in the Milky Way more than a hundred years ago.

What Causes a Supernova?

A supernova happens where there is a change in the core, or center, of a star. A change can occur in two different ways, with both resulting in a supernova.

The first type of supernova happens in binary star systems. Binary stars are two stars that orbit the same point. One of the stars, a carbon-oxygen white dwarf, steals matter from its companion star. Eventually, the white dwarf accumulates too much matter. Having too much matter causes the star to explode, resulting in a supernova.

The second type of supernova occurs at the end of a single star's lifetime. As the star runs out of nuclear fuel, some of its mass flows into its core. Eventually, the core is so heavy that it cannot withstand its own gravitational force. The core collapses, which results in the giant explosion of a supernova. The sun is a single star, but it does not have enough mass to become a supernova.

On the left is Supernova 1987A after the star has exploded. On the right is the star before it exploded.

Why do Scientists Study Supernovas?

A supernova burns for only a short period of time, but it can tell scientists a lot about the universe. One kind of supernova has shown scientists that we live in an expanding universe, one that is growing at an ever increasing rate.

Scientists also have determined that supernovas play a key role in distributing elements throughout the universe. When the star explodes, it shoots elements and debris into space. Many of the elements we find here on Earth are made in the core of stars. These elements travel on to form new stars, planets and everything else in the universe.

How do NASA Scientists Look for Supernovas?

NASA scientists use different types of telescopes to look for and study supernovas. Some telescopes are used to observe the visible light from the explosion. Others record data from the X-rays and gamma rays that are also produced. Both NASA's Hubble Space Telescope and Chandra X-ray Observatory have captured images of supernovas.

In June 2012, NASA launched the first orbiting telescope that focuses light in the high-energy region of the electromagnetic spectrum. The NuSTAR mission has a number of jobs to do. It will look for collapsed stars and black holes. It also will search for the remains of supernovas. Scientists hope to learn more about how stars explode and the elements that are created by supernovas.

References

- What-is-astronomy: telescopeobserver.com, Retrieved 21 April, 2020
- What-is-astronomy: thoughtco.com, Retrieved 15 January, 2020
- Celestial-bodies, physics: vedantu.com, Retrieved 21 July, 2020
- What-is-a-supernova, nasa-knows, features, for students, audience: nasa.gov, Retrieved 7 May, 2020

Sub-disciplines of Astronomy

CHAPTER 2

There are many sub-disciplines that fall under astronomy such as radio astronomy, exoplanetology, astrobiology, infrared astronomy, visible-light astronomy, ultraviolet astronomy, gamma-ray astronomy, etc. This chapter has been carefully written to provide an easy understanding of the various sub-disciplines of astronomy.

RADIO ASTRONOMY

Astronomical research is almost solely dependent on information we gather from the light of celestial bodies — their brightness, color, and movement. For millennia, humans could only observe the cosmos in visible light, but visible light constitutes only a tiny portion of the universe's radiation. The electromagnetic spectrum contains all radiation, from the longest wavelengths (lowest frequencies) to the shortest wavelengths (highest frequencies). Longer wavelengths correspond to lower-energy photons, and the lowest of these are radio waves. Radio astronomy is the study of the universe through analysis of radio emission from celestial objects.

Radio waves cut more easily through interstellar gas and dust, because they are not as easily scattered or absorbed as other, shorter-wavelength radiation. Some particularly strong sources of radio emission in the universe include active galactic nuclei (thought to be powered by supermassive black holes) and spinning stellar corpses known as pulsars. Closer to home, ham radio operators can most easily pick up the Sun and Jupiter, and with some fine-tuning, meteor showers.

Just as optical telescopes collect visible light, bring it to a focus, amplify it and make it available for analysis by various instruments, so do radio telescopes collect weak radio light waves, bring it to a focus, amplify it and make it available for analysis. We use radio telescopes to study naturally occurring radio light from stars, galaxies, black holes, and other astronomical objects. We can also use them to transmit and reflect radio light off of planetary bodies in our solar system. These specially-designed telescopes observe the longest wavelengths of light, ranging from 1 millimeter to over 10 meters long. For comparison, visible light waves are only a few hundred nanometers long, and a nanometer is only 1/10,000th the thickness of a piece of paper. In fact, we don't usually refer to radio light by its wavelength, but by its frequency.

Naturally occurring radio waves are extremely weak by the time they reach us from space. A cell phone signal is a billion billion times more powerful than the cosmic waves our telescopes detect.

Parts of a Radio Telescope

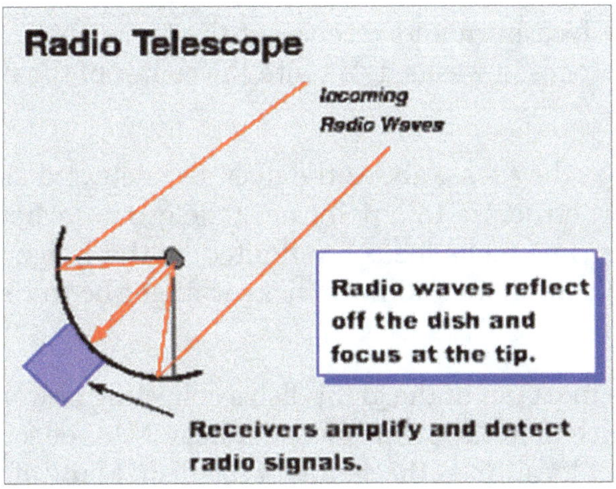

Radio telescopes are built in all shapes and sizes based on the kind of radio waves they pick up. However, every radio telescope has an antenna on a mount and at least one piece of receiver equipment to detect the signals.

Because radio waves are so long and cosmic radio sources are extremely weak, radio telescopes are the largest telescopes in the world, and only the most sensitive radio receivers are used inside them. Unfortunately, these huge antennas also pick up radio interference from modern electronics, and great effort is taken to protect radio telescopes from radio frequency interference.

The Antenna

The most basic antenna is a metal dipole antenna, often used on cars to pick up the radio waves broadcasters use to carry their audio shows.

The most versatile and powerful type of radio telescope is the parabolic dish antenna. The parabola is a useful mathematical shape that forces incoming radio waves to bounce up to a single point above it, called a focus. Dish antennae bounce many different wavelengths at once, and we need different receivers to tune to different frequency channels for the different kinds of research we do. To observe a specific wavelength range, we select a specific size funnel to grab the radio waves we want. These funnels are called feed horns, and our largest is the size of a pickup truck.

The diameter of the narrow end of each feed horn is the same size as a critical wavelength of the channel we want. What that means is that when the specific radio wave travels to the narrow end of its particular horn, it is beating perfectly against the sides, and the horn becomes the true antenna detecting the pulse. Here, we place a super-cooled receiver to collect the back and forth pulse of the wave as a signal it can send to the computer.

Feed Supports

We can either hang a feed horn and receiver at the focus above the dish, or install a mirror to redirect the focused waves down into the center of the dish where we can set multiple receivers.

If we place receivers at the focus, above the dish, the detected signal travels by cable along the feed support structure to a point near the ground where it can be recorded and analyzed. These prime focus feeds are limited by the weight and size of the feed horn that will safely fit up there and how tricky it might be to reach them for human maintenance.

More often, to get the most out of the giant dish's collecting power, we use a secondary mirror called a subreflector at the prime focus (or near it) to reflect focused waves down into a more convenient location — the center of the dish. Many of the subreflectors can be tilted to aim at the different feed horns in the center of the dish or to catch a glancing view of the sky to gather data about air quality conditions.

Dish Surface

If the lengths of the radio waves we're studying are very small, such as the millimeter waves collected by ALMA, then the perfection of the telescope's dish surface is critical. Any warp, bump, or ding in the parabola will scatter these tiny waves away from the focus, and we'll lose information.

We also have to consider the extreme environments where radio telescopes may operate. Wind and temperature differences can deform the parabola of a big radio telescope's dish and the pull of gravity affects the heavy antenna as it tilts to different parts of the sky. Therefore, the dishes of ALMA are kept small in order to better control their perfect shapes under these constantly varying conditions.

If the size of the radio wavelength being observed is very long, such as the centimeter waves picked up by the VLA and the VLBA, then the perfection of the dish's shape is not as critical to keep excellent observations of the radio sky. Those dishes are made rigid and tough and withstand the rigors of moving and working in various conditions.

Telescope Mounts

The dishes of some radio telescopes spin around a shaft that is aimed at the North Pole Star. These equatorial mounts allow the telescope to follow a position in the sky as the Earth rotates, simply by copying the Earth's axis of rotation and moving against it. But large equatorially-mounted radio telescopes are difficult to build, because they require millions of pounds of telescope to balance at many awkward angles. The largest ever built is our 140-foot (43-meter) dish telescope in Green Bank.

In most modern radio telescopes, a digital computer drives the telescope on simpler tilt and turn axes. This is how we can fully-steer 17 millions pounds of the GBT all across the sky.

Data Processing

In early radio telescopes, we had to tune into single, specific frequencies to watch for signals molecules of gas in space. Modern radio telescopes observe a large number of frequencies all at once, with computers dividing the frequency band into as many as several thousand separate channels that may range over tens to hundreds of megahertz. This innovation has changed radio telescopes from the equivalent of black and white cameras to full color.

In order to detect the faintest signals, the telescope remains staring at its radio source for hours, similar to keeping the shutter of a camera open. Our computer software keeps adding the waves together repeatedly to increase the signals from astronomical phenomena, and let the random noise signals coming from the receiver and the Earth's atmosphere average out over time.

Why we use Arrays?

Flying southward over the Very Large Array in central New Mexico shows the flatness of the Plains of San Agustin where the array is sited. At over 7,000-feet elevation, this ancient lakebed provides the space needed to lay 40 miles of double-track rails required to extend the VLA's Y-shape to give us its highest resolution capability.

The ability of a radio telescope to distinguish fine detail in the sky, called angular resolution, depends on the wavelength of observations divided by the size of the antenna. In other words, to get finer detailed views of the sky, the result of that simple equation needs to be a very small number.

Radio telescopes observe long wavelengths, so even when we divide our shortest radio wavelengths by our largest antennas, we still only have an angular resolution similar to that of your unaided eye observing the sky. To have their resolution compare to optical telescopes, a radio telescope's antenna size needs to be much, much larger.

The largest moving radio dish is the Green Bank Telescope, 100 meters across and fully-steerable. And that's about the maximum size for safely and accurately controlling a moving radio dish. The world's most gargantuan radio dish, the 1000-foot bowl in Arecibo, Puerto Rico, cannot move, but it can point on the sky by moving its receivers. The dish is supported inside a large sinkhole in the island's karst terrain.

Math finally cracked the conundrum: combine the views of a group of antennas spread over a large area to operate together as one gigantic telescope. This innovation won a Nobel Prize in physics.

Here's how it works: Two radio telescopes observe the same radio source. The telescopes are a known distance apart on the ground. The radio waves coming from the source will therefore arrive at one telescope at a slightly different time than the other. We see the slight difference in the way the waves show up, with one arriving slightly behind the other. The difference is a time delay in the phase of the wave.

The Atacama Large Millimeter/submillimeter Array links its 66 near-mirror reflecting dishes with a supercomputer.

When we combine the two offset waves, they will not overlap perfectly due to their phase shift, creating what we call interference fringes. As the Earth turns and the telescopes tilt to keep watching their source setting, the angles of their observations change. This tracking movement of the telescopes changes the distances the radio light travels from the source to each of the telescopes, in the same way that shadows are longer when the Sun is lower. This translates to different phase delays between the waves reaching each telescope. The longer we observe, the more variations we get. The more variations we get, the more perspectives we have on the object we're observing. And the farther apart we separate the telescopes, the sharper their binocular view of the sky becomes.

In arrays such as the VLA and ALMA, a device called a central local oscillator sends a common, lower frequency timing reference signal down fiber-optic cables to each antenna, behaving much like a conductor keeping a manageable tempo for the orchestra. The data received by each antenna are mixed with the local oscillator signal and then travel back down the fiber, to the main computer known as the correlator. The lower frequency signal allows the correlator to process and combine the data from each telescope at a rate that the computers can handle. However, the telescope arrays still need some of the most advanced computing technology in the world to handle the data.

The correlator synchronizes the incoming data from the different antennas to within a few millionths of a second of each other. It pairs up each antenna to every other antenna in the array, creating hundreds of unique perspectives on the same object. For every minute of observations, the perspectives change. To keep up with this constant and complex data stream, our correlators are among the fastest supercomputers in the world, performing their calculations at femtosecond speeds – up to 16 quadrillion operations every second.

Special software designed by radio astronomers and software engineers then assembles the data to create maps of radio objects in the sky.

SUBMILLIMETRE ASTRONOMY

Submillimetre astronomy or submillimeter astronomy is the branch of observational astronomy that is conducted at submillimetre wavelengths (i.e., terahertz radiation) of the electromagnetic spectrum. Astronomers place the submillimetre waveband between the far-infrared and microwave wavebands, typically taken to be between a few hundred micrometres and a millimetre. It is still common in submillimetre astronomy to quote wavelengths in 'microns', the old name for micrometre.

Using submillimetre observations, astronomers examine molecular clouds and dark cloud cores with a goal of clarifying the process of star formation from earliest collapse to stellar birth. Submillimetre observations of these dark clouds can be used to determine chemical abundances and cooling mechanisms for the molecules which comprise them. In addition, submillimetre observations give information on the mechanisms for the formation and evolution of galaxies.

Submillimetre Astronomy from the Ground

Panoramic view of the Chajnantor plateau, spanning about 180 degrees from north (on the left) to south (on the right) shows the antennas of the Atacama Large Millimeter Array.

The most significant limitation to the detection of astronomical emission at submillimetre wavelengths with ground-based observatories is atmospheric emission, noise and attenuation. Like the infrared, the submillimetre atmosphere is dominated by numerous water vapour absorption bands and it is only through "windows" between these bands that observations are possible. The ideal submillimetre observing site is dry, cool, has stable weather conditions and is away from urban population centres. There are only a handful of such sites identified, they include Mauna Kea (Hawaii, United States), the Llano de Chajnantor Observatory on the Atacama Plateau (Chile), the South Pole, and Hanle in India (the Himalayan site of the Indian Astronomical Observatory). Comparisons show that all four sites are excellent for submillimetre astronomy, and of these sites Mauna Kea is the most established and arguably the most accessible. There has been some recent interest in high-altitude Arctic sites, particularly Summit Station in Greenland where the PWV (precipitable water vapor) measure is always better than at Mauna Kea (however Mauna Kea's equatorial latitude of 19 degrees means it can observe more of the southern skies than Greenland).

The Llano de Chajnantor Observatory site hosts the Atacama Pathfinder Experiment (APEX), the largest submillimetre telescope operating in the southern hemisphere, and the world's largest ground based astronomy project, the Atacama Large Millimeter Array (ALMA), an interferometer for submillimetre wavelength observations made of 54 12-metre and 12 7-metre radio telescopes. The Submillimeter Array (SMA) is another interferometer, located at Mauna Kea, consisting of eight 6-metre diameter radio telescopes. The largest existing submillimetre telescope, the James Clerk Maxwell Telescope, is also located on Mauna Kea.

Submillimetre Astronomy from Near-space

With high-altitude balloons and aircraft, one can get above even more of the atmosphere. The BLAST experiment and SOFIA are two examples, respectively, although SOFIA can also handle near infrared observations.

Submillimetre Astronomy from Space

Comparison			
Name	Year	Wavelength	Aperture
Human Eye	-	0.39-0.75 μm	0.01 m
SWAS	1998	540 - 610 μm	0.55 - 0.7
Herschel	2009	55-672 μm	3.5 m

Space-based observations at the submillimetre wavelengths remove the ground-based limitations of atmospheric absorption. The Submillimeter Wave Astronomy Satellite (SWAS) was launched into low Earth orbit on December 5, 1998 as one of NASA's Small Explorer Program (SMEX) missions. The mission of the spacecraft is to make targeted observations of giant molecular clouds and dark cloud cores. The focus of SWAS is five spectral lines: water (H_2O), isotopic water ($H_2{}^{18}O$), isotopic carbon monoxide (^{13}CO), molecular oxygen (O_2), and neutral carbon (C I).

The SWAS satellite was repurposed in June, 2005 to provide support for the NASA *Deep Impact* mission. SWAS provided water production data on the comet until the end of August 2005.

The European Space Agency launched a space-based mission known as the Herschel Space Observatory (formerly called Far Infrared and Sub-millimetre Telescope or FIRST) in 2009. Herschel deployed the largest mirror ever launched into space and studied radiation in the far infrared and submillimetre wavebands. Rather than an Earth orbit, Herschel entered into a Lissajous orbit around L_2, the second Lagrangian point of the Earth-Sun system. L_2 is located approximately 1.5 million km from Earth and the placement of Herschel there lessened the interference by infrared and visible radiation from the Earth and Sun. Herschel's mission focused primarily on the origins of galaxies and galactic formation.

ASTROPHYSICS

Astrophysics is the branch of astronomy that employs the principles of physics and chemistry "to ascertain the nature of the astronomical objects, rather than their positions or motions in space". Among the objects studied are the Sun, other stars, galaxies, extrasolar planets, the interstellar medium and the cosmic microwave background. Emissions from these objects are examined across all parts of the electromagnetic spectrum, and the properties examined include luminosity, density, temperature, and chemical composition. Because astrophysics is a very broad subject, astrophysicists apply concepts and methods from many disciplines of physics, including classical mechanics, electromagnetism, statistical mechanics, thermodynamics, quantum mechanics, relativity, nuclear and particle physics, and atomic and molecular physics.

In practice, modern astronomical research often involves a substantial amount of work in the realms of theoretical and observational physics. Some areas of study for astrophysicists include their attempts to determine the properties of dark matter, dark energy, black holes, and other celestial bodies; whether or not time travel is possible, wormholes can form, or the multiverse exists; and the origin and ultimate fate of the universe. Topics also studied by theoretical astrophysicists include Solar System formation and evolution; stellar dynamics and evolution; galaxy formation and evolution; magnetohydrodynamics; large-scale structure of matter in the universe; origin of cosmic rays; general relativity, special relativity, quantum and physical cosmology, including string cosmology and astroparticle physics.

Astronomy is an ancient science, long separated from the study of terrestrial physics. In the Aristotelian worldview, bodies in the sky appeared to be unchanging spheres whose only motion was uniform motion in a circle, while the earthly world was the realm which underwent growth and decay and in which natural motion was in a straight line and ended when the moving object reached its goal. Consequently, it was held that the celestial region was made of a fundamentally different kind of matter from that found in the terrestrial sphere; either Fire as maintained by Plato, or Aether as maintained by Aristotle. During the 17th century, natural philosophers such as Galileo, Descartes, and Newton began to maintain that the celestial and terrestrial regions were made of similar kinds of material and were subject to the same natural laws. Their challenge was that the tools had not yet been invented with which to prove these assertions.

For much of the nineteenth century, astronomical research was focused on the routine work of measuring the positions and computing the motions of astronomical objects. A new astronomy, soon to be called astrophysics, began to emerge when William Hyde Wollaston and Joseph von Fraunhofer independently discovered that, when decomposing the light from the Sun, a multitude of dark lines (regions where there was less or no light) were observed in the spectrum. By 1860 the physicist, Gustav Kirchhoff, and the chemist, Robert Bunsen, had demonstrated that the dark lines in the solar spectrum corresponded to bright lines in the spectra of known gases, specific lines corresponding to unique chemical elements. Kirchhoff deduced that the dark lines in the solar spectrum are caused by absorption by chemical elements in the Solar atmosphere. In this way it was proved that the chemical elements found in the Sun and stars were also found on Earth.

Among those who extended the study of solar and stellar spectra was Norman Lockyer, who in 1868 detected radiant, as well as dark, lines in solar spectra. Working with the chemist, Edward Frankland, to investigate the spectra of elements at various temperatures and pressures, he could not associate a yellow line in the solar spectrum with any known elements. He thus claimed the line represented a new element, which was called helium, after the Greek Helios, the Sun personified.

In 1885, Edward C. Pickering undertook an ambitious program of stellar spectral classification at Harvard College Observatory, in which a team of woman computers, notably Williamina Fleming, Antonia Maury, and Annie Jump Cannon, classified the spectra recorded on photographic plates. By 1890, a catalog of over 10,000 stars had been prepared that grouped them into thirteen spectral types. Following Pickering's vision, by 1924 Cannon expanded the catalog to nine volumes and over a quarter of a million stars, developing the Harvard Classification Scheme which was accepted for worldwide use in 1922.

In 1895, George Ellery Hale and James E. Keeler, along with a group of ten associate editors from Europe and the United States, established The Astrophysical Journal: An International Review of Spectroscopy and Astronomical Physics. It was intended that the journal would fill the gap between journals in astronomy and physics, providing a venue for publication of articles on astronomical applications of the spectroscope; on laboratory research closely allied to astronomical physics, including wavelength determinations of metallic and gaseous spectra and experiments on radiation and absorption; on theories of the Sun, Moon, planets, comets, meteors, and nebulae; and on instrumentation for telescopes and laboratories.

Around 1920, following the discovery of the Hertzsprung–Russell diagram still used as the basis for classifying stars and their evolution, Arthur Eddington anticipated the discovery and mechanism of nuclear fusion processes in stars, in his paper The Internal Constitution of the Stars. At that time, the source of stellar energy was a complete mystery; Eddington correctly speculated that the source was fusion of hydrogen into helium, liberating enormous energy according to Einstein›s equation E = mc2. This was a particularly remarkable development since at that time fusion and thermonuclear energy, and even that stars are largely composed of hydrogen, had not yet been discovered.

In 1925 Cecilia Helena Payne (later Cecilia Payne-Gaposchkin) wrote an influential doctoral dissertation at Radcliffe College, in which she applied ionization theory to stellar atmospheres to relate the spectral classes to the temperature of stars. Most significantly, she discovered that hydrogen and helium were the principal components of stars. Despite Eddington's suggestion, this discovery was so unexpected that her dissertation readers convinced her to modify the conclusion before publication. However, later research confirmed her discovery.

By the end of the 20th century, studies of astronomical spectra had expanded to cover wavelengths extending from radio waves through optical, x-ray, and gamma wavelengths. In the 21st century it further expanded to include observations based on gravitational waves.

Observational Astrophysics

Observational astronomy is a division of the astronomical science that is concerned with recording and interpreting data, in contrast with theoretical astrophysics, which is mainly concerned with finding out the measurable implications of physical models. It is the practice of observing celestial objects by using telescopes and other astronomical apparatus.

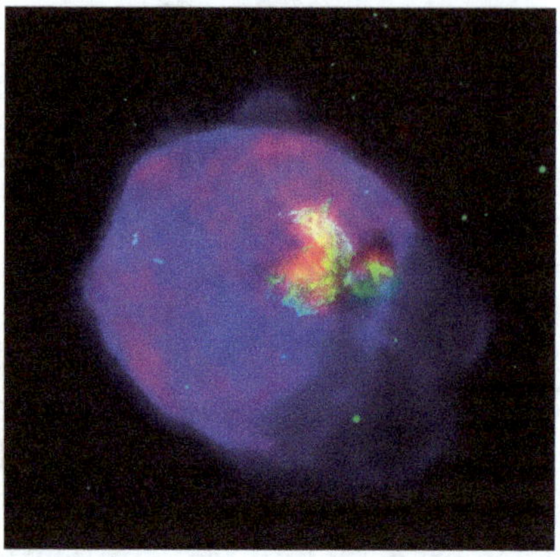

Supernova remnant LMC N 63A imaged in x-ray (blue), optical (green) and radio (red) wavelengths. The X-ray glow is from material heated to about ten million degrees Celsius by a shock wave generated by the supernova explosion.

The majority of astrophysical observations are made using the electromagnetic spectrum.

- Radio astronomy studies radiation with a wavelength greater than a few millimeters. Example areas of study are radio waves, usually emitted by cold objects such as interstellar gas and dust clouds; the cosmic microwave background radiation which is the redshifted light from the Big Bang; pulsars, which were first detected at microwave frequencies. The study of these waves requires very large radio telescopes.

- Infrared astronomy studies radiation with a wavelength that is too long to be visible to the naked eye but is shorter than radio waves. Infrared observations are usually made with telescopes similar to the familiar optical telescopes. Objects colder than stars (such as planets) are normally studied at infrared frequencies.

- Optical astronomy was the earliest kind of astronomy. Telescopes paired with a charge-coupled device or spectroscopes are the most common instruments used. The Earth›s atmosphere interferes somewhat with optical observations, so adaptive optics and space telescopes are used to obtain the highest possible image quality. In this wavelength range, stars are highly visible, and many chemical spectra can be observed to study the chemical composition of stars, galaxies and nebulae.

- Ultraviolet, X-ray and gamma ray astronomy study very energetic processes such as binary pulsars, black holes, magnetars, and many others. These kinds of radiation do not penetrate the Earth's atmosphere well. There are two methods in use to observe this part of the electromagnetic spectrum—space-based telescopes and ground-based imaging air Cherenkov telescopes (IACT). Examples of Observatories of the first type are RXTE, the Chandra X-ray Observatory and the Compton Gamma Ray Observatory. Examples of IACTs are the High Energy Stereoscopic System (H.E.S.S.) and the MAGIC telescope.

Other than electromagnetic radiation, few things may be observed from the Earth that originate from great distances. A few gravitational wave observatories have been constructed, but gravitational waves are extremely difficult to detect. Neutrino observatories have also been built, primarily to study our Sun. Cosmic rays consisting of very high energy particles can be observed hitting the Earth's atmosphere.

Observations can also vary in their time scale. Most optical observations take minutes to hours, so phenomena that change faster than this cannot readily be observed. However, historical data on some objects is available, spanning centuries or millennia. On the other hand, radio observations may look at events on a millisecond timescale (millisecond pulsars) or combine years of data (pulsar deceleration studies). The information obtained from these different timescales is very different.

The study of our very own Sun has a special place in observational astrophysics. Due to the tremendous distance of all other stars, the Sun can be observed in a kind of detail unparalleled by any other star. Our understanding of our own Sun serves as a guide to our understanding of other stars.

The topic of how stars change, or stellar evolution, is often modeled by placing the varieties of star types in their respective positions on the Hertzsprung–Russell diagram, which can be viewed as representing the state of a stellar object, from birth to destruction.

Theoretical Astrophysics

Stream lines on this simulation of a supernova show the flow of matter behind the shock wave giving clues as to the origin of pulsars

Theoretical astrophysicists use a wide variety of tools which include analytical models (for example, polytropes to approximate the behaviors of a star) and computational numerical simulations. Each has some advantages. Analytical models of a process are generally better for giving insight into the heart of what is going on. Numerical models can reveal the existence of phenomena and effects that would otherwise not be seen.

Theorists in astrophysics endeavor to create theoretical models and figure out the observational consequences of those models. This helps allow observers to look for data that can refute a model or help in choosing between several alternate or conflicting models.

Theorists also try to generate or modify models to take into account new data. In the case of an inconsistency, the general tendency is to try to make minimal modifications to the model to fit the data. In some cases, a large amount of inconsistent data over time may lead to total abandonment of a model.

Topics studied by theoretical astrophysicists include stellar dynamics and evolution; galaxy formation and evolution; magnetohydrodynamics; large-scale structure of matter in the universe; origin of cosmic rays; general relativity and physical cosmology, including string cosmology and astroparticle physics. Astrophysical relativity serves as a tool to gauge the properties of large scale structures for which gravitation plays a significant role in physical phenomena investigated and as the basis for black hole (astro) physics and the study of gravitational waves.

Some widely accepted and studied theories and models in astrophysics, now included in the Lambda-CDM model, are the Big Bang, cosmic inflation, dark matter, dark energy and fundamental theories of physics. Wormholes are examples of hypotheses which are yet to be proven (or disproven).

Popularization

The roots of astrophysics can be found in the seventeenth century emergence of a unified physics, in which the same laws applied to the celestial and terrestrial realms. There were scientists who were qualified in both physics and astronomy who laid the firm foundation for the current science of astrophysics. In modern times, students continue to be drawn to astrophysics due to its popularization by the Royal Astronomical Society and notable educators such as prominent professors Lawrence Krauss, Subrahmanyan Chandrasekhar, Stephen Hawking, Hubert Reeves, Carl Sagan, Neil deGrasse Tyson and Patrick Moore. The efforts of the early, late, and present scientists continue to attract young people to study the history and science of astrophysics.

ASTROBIOLOGY

Astrobiology, formerly known as exobiology, is an interdisciplinary scientific field concerned with the origins, early evolution, distribution, and future of life in the universe. Astrobiology considers the question of whether extraterrestrial life exists, and if it does, how humans can detect it.

Astrobiology makes use of molecular biology, biophysics, biochemistry, chemistry, astronomy, physical cosmology, exoplanetology and geology to investigate the possibility of life on other worlds and help recognize biospheres that might be different from that on Earth. The origin and early evolution of life is an inseparable part of the discipline of astrobiology. Astrobiology concerns itself with interpretation of existing scientific data, and although speculation is entertained to give context, astrobiology concerns itself primarily with hypotheses that fit firmly into existing scientific theories.

This interdisciplinary field encompasses research on the origin of planetary systems, origins of organic compounds in space, rock-water-carbon interactions, abiogenesis on Earth, planetary habitability, research on biosignatures for life detection, and studies on the potential for life to adapt to challenges on Earth and in outer space.

Biochemistry may have begun shortly after the Big Bang, 13.8 billion years ago, during a habitable epoch when the Universe was only 10–17 million years old. According to the panspermia hypothesis, microscopic life—distributed by meteoroids, asteroids and other small Solar System bodies—may exist throughout the universe. According to research published in August 2015, very large galaxies may be more favorable to the creation and development of habitable planets than such smaller galaxies as the Milky Way. Nonetheless, Earth is the only place in the universe humans know to harbor life. Estimates of habitable zones around other stars, sometimes referred to as "Goldilocks zones," along with the discovery of hundreds of extrasolar planets and new insights into extreme habitats here on Earth, suggest that there may be many more habitable places in the universe than considered possible until very recently.

Current studies on the planet Mars by the Curiosity and Opportunity rovers are searching for evidence of ancient life as well as plains related to ancient rivers or lakes that may have been habitable. The search for evidence of habitability, taphonomy (related to fossils), and organic molecules on the planet Mars is now a primary NASA and ESA objective.

Even if extraterrestrial life is never discovered, the interdisciplinary nature of astrobiology, and the cosmic and evolutionary perspectives engendered by it, may still result in a range of benefits here on Earth.

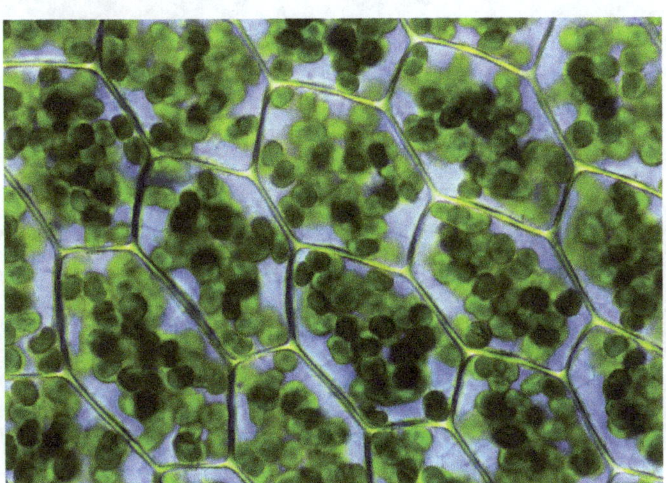

It is not known whether life elsewhere in the universe would utilize cell structures like those found on Earth.

The term was first proposed by the Russian astronomer Gavriil Tikhov in 1953. The synonyms of astrobiology are diverse; however, the synonyms were structured in relation to the most important sciences implied in its development: astronomy and biology. The term exobiology was coined by molecular biologist and Nobel Prize winner Joshua Lederberg. Exobiology is considered to have a narrow scope limited to search of life external to Earth, whereas subject area of astrobiology is wider and investigates the link between life and the universe, which includes the search for extraterrestrial life, but also includes the study of life on Earth, its origin, evolution and limits.

The term xenobiology is now used in a more specialized sense, to mean "biology based on foreign chemistry", whether of extraterrestrial or terrestrial (possibly synthetic) origin. Since alternate chemistry analogs to some life-processes have been created in the laboratory, xenobiology is now considered as an extant subject.

While it is an emerging and developing field, the question of whether life exists elsewhere in the universe is a verifiable hypothesis and thus a valid line of scientific inquiry. Though once considered outside the mainstream of scientific inquiry, astrobiology has become a formalized field of study. Planetary scientist David Grinspoon calls astrobiology a field of natural philosophy, grounding speculation on the unknown, in known scientific theory. NASA's interest in exobiology first began with the development of the U.S. Space Program. In 1959, NASA funded its first exobiology project, and in 1960, NASA founded an Exobiology Program, which is now one of four main elements of NASA's current Astrobiology Program. In 1971, NASA funded the search for extraterrestrial intelligence (SETI) to search radio frequencies of the electromagnetic spectrum for interstellar communications transmitted by extraterrestrial life outside the Solar System. NASA's Viking missions to Mars, launched in 1976, included three biology experiments designed to look for metabolism of present life on Mars.

In June 2014, the John W. Kluge Center of the Library of Congress held a seminar focusing on astrobiology. Panel members (L to R) Robin Lovin, Derek Malone-France, and Steven J. Dick.

Advancements in the fields of astrobiology, observational astronomy and discovery of large varieties of extremophiles with extraordinary capability to thrive in the harshest environments on Earth, have led to speculation that life may possibly be thriving on many of the extraterrestrial bodies in the universe. A particular focus of current astrobiology research is the search for life on Mars due to this planet's proximity to Earth and geological history. There is a growing body of evidence to suggest that Mars has previously had a considerable amount of water on its surface, water being considered an essential precursor to the development of carbon-based life.

Missions specifically designed to search for current life on Mars were the Viking program and Beagle 2 probes. The Viking results were inconclusive, and Beagle 2 failed minutes after landing. A future mission with a strong astrobiology role would have been the Jupiter Icy Moons Orbiter, designed to study the frozen moons of Jupiter—some of which may have liquid water—had it not been cancelled. In late 2008, the Phoenix lander probed the environment for past and present planetary habitability of microbial life on Mars, and researched the history of water there.

The European Space Agency's astrobiology roadmap from 2016, identified five main research topics, and specifies several key scientific objectives for each topic. The five research topics are: 1) Origin and evolution of planetary systems; 2) Origins of organic compounds in space; 3) Rock-water-carbon interactions, organic synthesis on Earth, and steps to life; 4) Life and habitability; 5) Biosignatures as facilitating life detection.

In November 2011, NASA launched the Mars Science Laboratory mission carrying the Curiosity rover, which landed on Mars at Gale Crater in August 2012. The Curiosity rover is currently probing the environment for past and present planetary habitability of microbial life on Mars. On 9 December 2013, NASA reported that, based on evidence from Curiosity studying Aeolis Palus, Gale Crater contained an ancient freshwater lake which could have been a hospitable environment for microbial life.

The European Space Agency is currently collaborating with the Russian Federal Space Agency (Roscosmos) and developing the ExoMars astrobiology rover, which is to be launched in July 2020. Meanwhile, NASA is developing the Mars 2020 astrobiology rover and sample cacher for a later return to Earth.

Methodology

Planetary Habitability

When looking for life on other planets like Earth, some simplifying assumptions are useful to reduce the size of the task of the astrobiologist. One is the informed assumption that the vast majority of life forms in our galaxy are based on carbon chemistries, as are all life forms on Earth. Carbon is well known for the unusually wide variety of molecules that can be formed around it. Carbon is the fourth most abundant element in the universe and the energy required to make or break a bond is at just the appropriate level for building molecules which are not only stable, but also reactive. The fact that carbon atoms bond readily to other carbon atoms allows for the building of extremely long and complex molecules.

The presence of liquid water is an assumed requirement, as it is a common molecule and provides an excellent environment for the formation of complicated carbon-based molecules that could eventually lead to the emergence of life. Some researchers posit environments of water-ammonia mixtures as possible solvents for hypothetical types of biochemistry.

A third assumption is to focus on planets orbiting Sun-like stars for increased probabilities of planetary habitability. Very large stars have relatively short lifetimes, meaning that life might not have time to emerge on planets orbiting them. Very small stars provide so little heat and warmth that only planets in very close orbits around them would not be frozen solid, and in such close orbits these planets would be tidally "locked" to the star. The long lifetimes of red dwarfs could allow the development of habitable environments on planets with thick atmospheres. This is significant, as red dwarfs are extremely common.

Since Earth is the only planet known to harbor life, there is no evident way to know if any of these simplifying assumptions are correct.

Communication Attempts

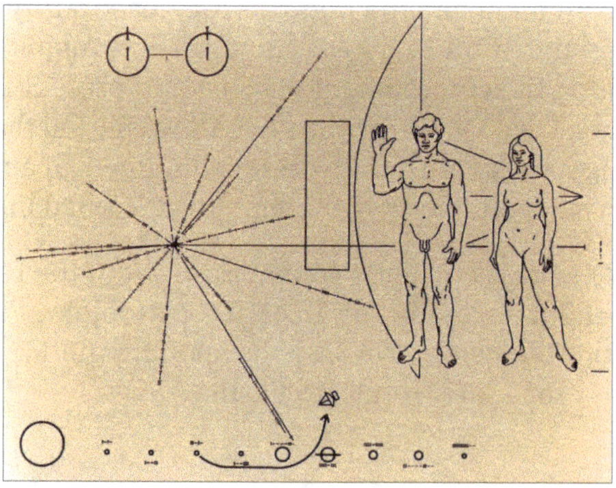

The illustration on the Pioneer plaque.

Research on communication with extraterrestrial intelligence (CETI) focuses on composing and deciphering messages that could theoretically be understood by another technological civilization. Communication attempts by humans have included broadcasting mathematical languages, pictorial systems such as the Arecibo message and computational approaches to detecting and deciphering 'natural' language communication. The SETI program, for example, uses both radio telescopes and optical telescopes to search for deliberate signals from an extraterrestrial intelligence.

While some high-profile scientists, such as Carl Sagan, have advocated the transmission of messages, scientist Stephen Hawking warned against it, suggesting that aliens might simply raid Earth for its resources and then move on.

Elements of Astrobiology

Astronomy

Artist's impression of the extrasolar planet OGLE-2005-BLG-390Lb orbiting its star 20,000 light-years from Earth; this planet was discovered with gravitational microlensing.

The NASA Kepler mission, launched in March 2009, searches for extrasolar planets.

Most astronomy-related astrobiology research falls into the category of extrasolar planet (exoplanet) detection, the hypothesis being that if life arose on Earth, then it could also arise on other planets with similar characteristics. To that end, a number of instruments designed to detect Earth-sized exoplanets have been considered, most notably NASA's Terrestrial Planet Finder (TPF) and ESA's Darwin programs, both of which have been cancelled. NASA launched the Kepler mission in March 2009, and the French Space Agency launched the COROT space mission in 2006. There are also several less ambitious ground-based efforts underway.

The goal of these missions is not only to detect Earth-sized planets, but also to directly detect light from the planet so that it may be studied spectroscopically. By examining planetary spectra, it would be possible to determine the basic composition of an extra-solar planet's atmosphere and/or surface. Given this knowledge, it may be possible to assess the likelihood of life being found on that planet. A NASA research group, the Virtual Planet Laboratory, is using computer modeling to generate a wide variety of virtual planets to see what they would look like if viewed by TPF or Darwin. It is hoped that once these missions come online, their spectra can be cross-checked with these virtual planetary spectra for features that might indicate the presence of life.

An estimate for the number of planets with intelligent communicative extraterrestrial life can be gleaned from the Drake equation, essentially an equation expressing the probability of intelligent life as the product of factors such as the fraction of planets that might be habitable and the fraction of planets on which life might arise:

$$N = R^{*} \times f_p \times n_e \times f_l \times f_i \times f_c \times L$$

where:

- N = The number of communicative civilizations,
- R^* = The rate of formation of suitable stars (stars such as our Sun),
- f_p = The fraction of those stars with planets (current evidence indicates that planetary systems may be common for stars like the Sun),
- n_e = The number of Earth-sized worlds per planetary system,
- f_l = The fraction of those Earth-sized planets where life actually develops,
- f_i = The fraction of life sites where intelligence develops,
- f_c = The fraction of communicative planets (those on which electromagnetic communications technology develops),
- L = The "lifetime" of communicating civilizations.

However, whilst the rationale behind the equation is sound, it is unlikely that the equation will be constrained to reasonable limits of error any time soon. The problem with the formula is that it is not usable to generate or support hypotheses because it contains factors that can never be verified. The first term, R^*, number of stars, is generally constrained within a few orders of magnitude. The second and third terms, f_p, stars with planets and f_e, planets with habitable conditions, are being evaluated for the star's neighborhood. Drake originally formulated the equation merely as an agenda for discussion at the Green Bank conference, but some applications of the formula had been taken literally and related to simplistic or pseudoscientific arguments. Another associated topic is the Fermi paradox, which suggests that if intelligent life is common in the universe, then there should be obvious signs of it.

Another active research area in astrobiology is planetary system formation. It has been suggested that the peculiarities of the Solar System (for example, the presence of Jupiter as a protective shield) may have greatly increased the probability of intelligent life arising on our planet.

Biology

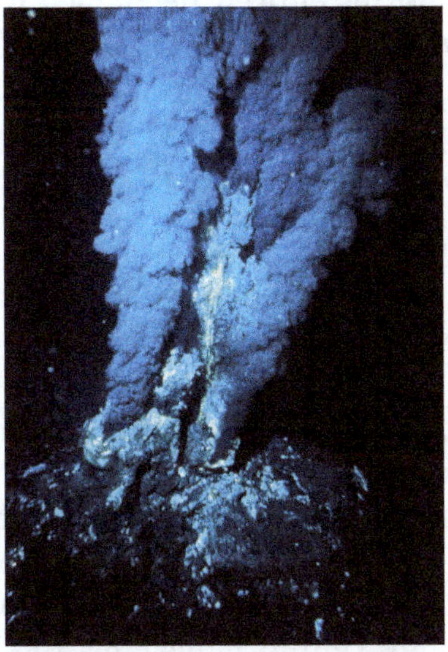

Hydrothermal vents are able to support extremophile bacteria on Earth and may also support life in other parts of the cosmos.

Biology cannot state that a process or phenomenon, by being mathematically possible, has to exist forcibly in an extraterrestrial body. Biologists specify what is speculative and what is not. The discovery of extremophiles, organisms able to survive in extreme environments, became a core research element for astrobiologists, as they are important to understand four areas in the limits of life in planetary context: the potential for panspermia, forward contamination due to human exploration ventures, planetary colonization by humans, and the exploration of extinct and extant extraterrestrial life.

Until the 1970s, life was thought to be entirely dependent on energy from the Sun. Plants on Earth's surface capture energy from sunlight to photosynthesize sugars from carbon dioxide and water, releasing oxygen in the process that is then consumed by oxygen-respiring organisms, passing their energy up the food chain. Even life in the ocean depths, where sunlight cannot reach, was thought to obtain its nourishment either from consuming organic detritus rained down from the surface waters or from eating animals that did. The world's ability to support life was thought to depend on its access to sunlight. However, in 1977, during an exploratory dive to the Galapagos Rift in the deep-sea exploration submersible Alvin, scientists discovered colonies of giant tube worms, clams, crustaceans, mussels, and other assorted creatures clustered around undersea volcanic features known as black smokers. These creatures thrive despite having no access to sunlight, and it was soon discovered that they comprise an entirely independent ecosystem. Although most of these multicellular lifeforms need dissolved oxygen (produced by oxygenic photosynthesis) for their aerobic cellular respiration and thus are not completely independent from sunlight by themselves, the basis for their food chain is a form of bacterium that derives its energy from oxidization of reactive chemicals, such as hydrogen or hydrogen sulfide, that bubble up from the Earth's interior. Other lifeforms entirely decoupled from the energy from sunlight are green sulphur bacteria which are capturing geothermal light for anoxygenic photosynthesis or bacteria running chemolithoautotrophy based on the radioactive decay of uranium. This chemosynthesis revolutionized the study of biology and astrobiology by revealing that life need not be sun-dependent; it only requires water and an energy gradient in order to exist.

Biologists have found extremophiles that thrive in ice, boiling water, acid, alkali, the water core of nuclear reactors, salt crystals, toxic waste and in a range of other extreme habitats that were previously thought to be inhospitable for life. This opened up a new avenue in astrobiology by massively expanding the number of possible extraterrestrial habitats. Characterization of these organisms, their environments and their evolutionary pathways, is considered a crucial component to understanding how life might evolve elsewhere in the universe. For example, some organisms able to withstand exposure to the vacuum and radiation of outer space include the lichen fungi Rhizocarpon geographicum and Xanthoria elegans, the bacterium Bacillus safensis, Deinococcus radiodurans, Bacillus subtilis, yeast Saccharomyces cerevisiae, seeds from Arabidopsis thaliana ('mouse-ear cress'), as well as the invertebrate animal Tardigrade. While tardigrades are not considered true extremophiles, they are considered extremotolerant microorganisms that have contributed to the field of astrobiology. Their extreme radiation tolerance and presence of DNA protection proteins may provide answers as to whether life can survive away from the protection of the Earth's atmosphere.

Jupiter's moon, Europa, and Saturn's moon, Enceladus, are now considered the most likely locations for extant extraterrestrial life in the Solar System due to their subsurface water oceans where radiogenic and tidal heating enables liquid water to exist.

The origin of life, known as abiogenesis, distinct from the evolution of life, is another ongoing field of research. Oparin and Haldane postulated that the conditions on the early Earth were conducive to the formation of organic compounds from inorganic elements and thus to the formation of many of the chemicals common to all forms of life we see today. The study of this process, known as prebiotic chemistry, has made some progress, but it is still unclear whether or not life could have formed in such a manner on Earth. The alternative hypothesis of panspermia is that the first elements of life may have formed on another planet with even more favorable conditions (or even in interstellar space, asteroids, etc.) and then have been carried over to Earth—the panspermia hypothesis.

The cosmic dust permeating the universe contains complex organic compounds ("amorphous organic solids with a mixed aromatic-aliphatic structure") that could be created naturally, and rapidly, by stars. Further, a scientist suggested that these compounds may have been related to the development of life on Earth and said that, "If this is the case, life on Earth may have had an easier time getting started as these organics can serve as basic ingredients for life."

More than 20% of the carbon in the universe may be associated with polycyclic aromatic hydrocarbons (PAHs), possible starting materials for the formation of life. PAHs seem to have been formed shortly after the Big Bang, are widespread throughout the universe, and are associated with new stars and exoplanets. PAHs are subjected to interstellar medium conditions and are transformed through hydrogenation, oxygenation and hydroxylation, to more complex organics – "a step along the path toward amino acids and nucleotides, the raw materials of proteins and DNA, respectively".

Astroecology

Astroecology concerns the interactions of life with space environments and resources, in planets, asteroids and comets. On a larger scale, astroecology concerns resources for life about stars in the galaxy through the cosmological future. Astroecology attempts to quantify future life in space, addressing this area of astrobiology.

Experimental astroecology investigates resources in planetary soils, using actual space materials in meteorites. The results suggest that Martian and carbonaceous chondrite materials can support bacteria, algae and plant (asparagus, potato) cultures, with high soil fertilities. The results support that life could have survived in early aqueous asteroids and on similar materials imported to Earth by dust, comets and meteorites, and that such asteroid materials can be used as soil for future space colonies.

On the largest scale, cosmoecology concerns life in the universe over cosmological times. The main sources of energy may be red giant stars and white and red dwarf stars, sustaining life for 1020 years. Astroecologists suggest that their mathematical models may quantify the potential amounts of future life in space, allowing a comparable expansion in biodiversity, potentially leading to diverse intelligent life forms.

Astrogeology

Astrogeology is a planetary science discipline concerned with the geology of celestial bodies such as the planets and their moons, asteroids, comets, and meteorites. The information gathered by this discipline allows the measure of a planet's or a natural satellite's potential to develop and sustain life, or planetary habitability.

An additional discipline of astrogeology is geochemistry, which involves study of the chemical composition of the Earth and other planets, chemical processes and reactions that govern the composition of rocks and soils, the cycles of matter and energy and their interaction with the hydrosphere and the atmosphere of the planet. Specializations include cosmochemistry, biochemistry and organic geochemistry.

The fossil record provides the oldest known evidence for life on Earth. By examining the fossil evidence, paleontologists are able to better understand the types of organisms that arose on the early Earth. Some regions on Earth, such as the Pilbara in Western Australia and the McMurdo Dry Valleys of Antarctica, are also considered to be geological analogs to regions of Mars, and as such, might be able to provide clues on how to search for past life on Mars.

The various organic functional groups, composed of hydrogen, oxygen, nitrogen, phosphorus, sulfur, and a host of metals, such as iron, magnesium, and zinc, provide the enormous diversity of chemical reactions necessarily catalyzed by a living organism. Silicon, in contrast, interacts with only a few other atoms, and the large silicon molecules are monotonous compared with the combinatorial universe of organic macromolecules. Indeed, it seems likely that the basic building blocks of life anywhere will be similar those on Earth, in the generality if not in the detail. Although terrestrial life and life that might arise independently of Earth are expected to use many similar, if not identical, building blocks, they also are expected to have some biochemical qualities that are unique. If life has had a comparable impact elsewhere in the Solar System, the relative abundances of chemicals key for its survival – whatever they may be – could betray its presence. Whatever extraterrestrial life may be, its tendency to chemically alter its environment might just give it away.

INFRARED ASTRONOMY

Infrared astronomy is the branch of astronomy and astrophysics that studies objects as they would appear in the infrared (IR) region of the electromagnetic spectrum. Visible light ranges in wavelength between 400 nanometers (nm) (blue) and 700 nm (red). Wavelengths that are longer than 700 nm but shorter than microwaves are grouped as infrared radiation (or submillimeter waves). Scientists classify infrared astronomy as part of optical astronomy because optical components (mirrors, lenses and solid state digital detectors) are usually used.

Infrared astronomy has helped scientists gain better insights into processes that occur during the birth and death of stars, as well as to observe brown dwarfs, extrasolar planets, and other sub-stellar mass objects. It has also aided the examination of dust grains in the Milky Way and other galaxies. The field of IR astronomy has greatly benefited from the development of highly sensitive IR detectors, the establishment of large arrays on the ground, and the use of satellite technology.

After Isaac Newton used prisms to split white light into a spectrum, William Herschel found in 1800 that the hottest part of the band of light from the Sun was actually past the red end of the spectrum. These "heat rays" were found to display some spectral lines. In 1856, Charles Piazzi Smyth detected infrared radiation in the light of the Moon.

An infrared image of the Seven Sisters (Pleiades Star Cluster).

Modern Infrared Astronomy

NASA's Spitzer Space Telescope, shown here prior to launch, is dedicated to infrared astronomy.

Near-infrared radiation (infrared radiation with wavelengths close to that of visible light) behaves in a manner very similar to visible light, and it can be detected using similar electronic devices. For this reason, the near-infrared region of the spectrum is commonly incorporated as part of the "optical" spectrum, along with the near-ultraviolet. (Most scientific instruments such as optical telescopes cover the near-infrared region as well as the visible region). The far infrared extends to submillimeter wavelengths, which are observed by telescopes such as the James Clerk Maxwell Telescope at Mauna Kea Observatory.

Like all other forms of electromagnetic radiation, infrared is utilized by astronomers to learn more about the universe. As infrared is essentially heat radiation, infrared telescopes (which include most major optical telescopes as well as a few dedicated infrared telescopes) need to have their detectors shielded from heat and chilled with liquid nitrogen in order to obtain images. This is particularly important in the mid-infrared and far-infrared regions of the spectrum.

The main limitation on infrared sensitivity from ground-based telescopes is that water vapor in the Earth's atmosphere absorbs a significant amount of infrared radiation. For this reason, most infrared telescopes are built in very dry places at high altitudes (above most of the water vapor in the atmosphere). Suitable locations on Earth include the Mauna Kea Observatory, situated at 4205 meters above sea level; the Atacama Large Millimeter Array (ALMA), at 5000 m in Chile; and high-altitude ice desert regions, such as Dome C in Antarctica.

However, as with visible-light telescopes, space is the ideal place for the use of IR-sensitive telescopes, and most optical telescopes launched into space (such as the Hubble Space Telescope) can also perform infrared observations. The Spitzer Space Telescope (formerly called the Space Infrared Telescope Facility, or SIRTF) is dedicated solely to infrared observations.

Another way of doing infrared astronomy is by the use of airborne observatories, such as SOFIA (Stratospheric Observatory for Infrared Astronomy) and the Kuiper Airborne Observatory. By flying at high altitude (in the stratosphere), there is less water vapor between the telescope and objects in space, leading to a smaller IR absorption of the atmosphere. The residual IR background (due to the remaining absorption) is statically removed by applying a chopping reduction technique of the observed field and a blank region.

Infrared Technology

The cheapest infrared detectors commonly used at research telescopes are arrays of HgCdTe (mercury, cadmium, and tellurium). These operate well at wavelengths between one and five micrometers. For longer wavelength observations or higher sensitivity, other detectors may be used, including other narrow gap semiconductor detectors, low temperature bolometer arrays, or photon-counting Superconducting Tunnel Junction arrays.

Special requirements for infrared astronomy include: very low, dark currents, to allow long integration times, associated low-noise readout circuits, and sometimes very high pixel counts.

Astronomers' Infrared Spectrum

Infrared space telescopes—such as the Spitzer Space Telescope, Infrared Astronomical Satellite (IRAS), Infrared Space Observatory (ISO), and the forthcoming Herschel Space Observatory—can observe across almost all of the infrared spectrum. However, most infrared astronomy continues to be carried out at ground-based telescopes, and these are limited to observations through a small number of spectral "windows," at wavelengths where the Earth's atmosphere is transparent. The main infrared windows are listed below:

Wavelength range (micrometres)	Astronomical bands	Telescopes
0.65 to 1.0	R and I bands	All major optical telescopes
1.25	J band	Most major optical telescopes and most dedicated infrared telescopes
1.65	H band	Most major optical telescopes and most dedicated infrared telescopes
2.2	K band	Most major optical telescopes and most dedicated infrared telescopes
3.45	L band	Most dedicated infrared telescopes and some optical telescopes
4.7	M band	Most dedicated infrared telescopes and some optical telescopes
10	N band	Most dedicated infrared telescopes and some optical telescopes
20	Q band	Some dedicated infrared telescopes and some optical telescopes
450	Submillimeter	Submillimeter telescopes

Between these windows, there are generally regions where infrared observations are more difficult or impossible to conduct from the ground, due to the opacity of the atmosphere. Dedicated infrared and submillimeter telescopes are generally built at very high altitude sites, such as the Mauna Kea Observatory in Hawaii and the ALMA site in Chile. Some telescopes are flown on aircraft like SOFIA, providing the best sensitivity available from Earth-based observatories.

Space Telescopes used for Infrared Astronomy

Infrared Astronomical Satellite (IRAS): Launched on January 25, 1983, IRAS was a joint project of the United States (NASA), the Netherlands (NIVR) and the United

Kingdom (SERC). It was the first-ever space-based observatory to perform a survey of the entire sky at infrared wavelengths. Its mission lasted ten months. It discovered about 350,000 sources, many of which still await identification.

- Infrared Space Observatory (ISO): This telescope was launched in November 1995 and operated until it ran out of liquid helium in May 1998. It was operated by the European Space Agency (ESA), in cooperation with ISAS (part of Japan Aerospace Exploration Agency, JAXA), and NASA. The ISO was designed to make observations in the infrared region of wavelengths between 2.5 and 240 micrometers.
- Spitzer Space Telescope: Launched on August 25, 2003, the Spitzer telescope is one of NASA's Great Observatories. It follows a heliocentric orbit (instead of a geocentric orbit). It follows Earth in its orbit, and drifts away from Earth at approximately 0.1 astronomical unit per year. The primary mirror is 85 cm in diameter, with a focal length that is 12 times the diameter of the primary mirror, and it is made of beryllium and cooled to 5.5 K.

EXOPLANETOLOGY

Exoplanetology, or exoplanetary science, is an integrated field of astronomical science dedicated to the search for and study of exoplanets (extrasolar planets). It employs an interdisciplinary approach which includes astrobiology, astrophysics, astronomy, astrochemistry, astrogeology, geochemistry, and planetary science.

Nomenclature

The exoplanet naming convention is an extension of the system used for naming multiple-star systems as adopted by the International Astronomical Union (IAU). For an exoplanet orbiting a single star, the name is normally formed by taking the name of its parent star and adding a lowercase letter. The first planet discovered in a system is given the designation "b" (the parent star is considered to be "a") and later planets are given subsequent letters. If several planets in the same system are discovered at the same time, the closest one to the star gets the next letter, followed by the other planets in order of orbit size. A provisional IAU-sanctioned standard exists to accommodate the naming of circumbinary planets. A limited number of exoplanets have IAU-sanctioned proper names. Other naming systems exist.

IAU

The official definition of "planet" used by the International Astronomical Union (IAU) only covers the Solar System and thus does not apply to exoplanets. As of April 2011, the only defining statement issued by the IAU that pertains to exoplanets is a working definition issued in 2001 and modified in 2003. That definition contains the following criteria:

- Objects with true masses below the limiting mass for thermonuclear fusion of deuterium (currently calculated to be 13 Jupiter masses for objects of solar metallicity) that orbit stars or stellar remnants are "planets" (no matter how they formed). The minimum mass/size required for an extrasolar object to be considered a planet should be the same as that used in the Solar System.
- Substellar objects with true masses above the limiting mass for thermonuclear fusion of deuterium are "brown dwarfs", no matter how they formed or where they are located.
- Free-floating objects in young star clusters with masses below the limiting mass for thermonuclear fusion of deuterium are not "planets", but are "sub-brown dwarfs" (or whatever name is most appropriate).

The IAU's working definition is not always used. One alternate suggestion is that planets should be distinguished from brown dwarfs on the basis of formation. It is widely thought that giant planets form through core accretion, which may sometimes produce planets with masses above the deuterium fusion threshold; massive planets of that sort may have already been observed. Brown dwarfs form like stars from the direct gravitational collapse of clouds of gas and this formation mechanism also produces objects that are below the 13 M_{Jup} limit and can be as low as 1 M_{Jup}. Objects in this mass range that orbit their stars with wide separations of hundreds or thousands of AU and have large star/object mass ratios likely formed as brown dwarfs; their atmospheres would likely have a composition more similar to their host star than accretion-formed planets which would contain increased abundances of heavier elements. Most directly imaged planets as of April 2014 are massive and have wide orbits so probably represent the low-mass end of brown dwarf formation. One study suggests that objects above 10 M_{Jup} formed through gravitational instability and should not be thought of as planets.

Also, the 13-Jupiter-mass cutoff does not have precise physical significance. Deuterium fusion can occur in some objects with a mass below that cutoff. The amount of deuterium fused depends to some extent on the composition of the object. As of 2011 the Extrasolar Planets Encyclopaedia included objects up to 25 Jupiter masses, saying, "The fact that there is no special feature around 13 M_{Jup} in the observed mass spectrum reinforces the choice to forget this mass limit". As of 2016 this limit was increased to 60 Jupiter masses based on a study of mass–density relationships. The Exoplanet Data Explorer includes objects up to 24 Jupiter masses with the advisory: "The 13 Jupiter-mass distinction by the IAU Working Group is physically unmotivated for planets with rocky cores, and observationally problematic due to the sin i ambiguity." The NASA Exoplanet Archive includes objects with a mass (or minimum mass) equal to or less than 30 Jupiter masses. Another criterion for separating planets and brown dwarfs, rather than deuterium fusion, formation process or location, is whether the core pressure is dominated by coulomb pressure or electron degeneracy pressure with the dividing line at around 5 Jupiter masses.

Detection Methods

Direct Imaging

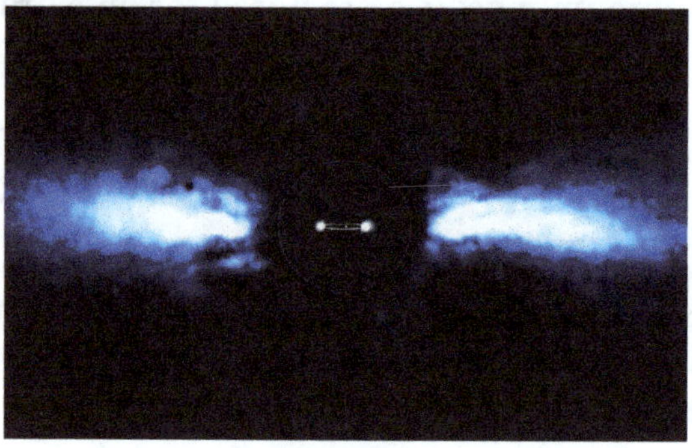

Directly imaged planet Beta Pictoris b.

Planets are extremely faint compared with their parent stars. For example, a Sun-like star is about a billion times brighter than the reflected light from any exoplanet orbiting it. It is difficult to detect such a faint light source, and furthermore the parent star causes a glare that tends to wash it out. It is necessary to block the light from the parent star in order to reduce the glare while leaving the light from the planet detectable; doing so is a major technical challenge which requires extreme optothermal stability. All exoplanets that have been directly imaged are both large (more massive than Jupiter) and widely separated from their parent star.

Specially designed direct-imaging instruments such as Gemini Planet Imager, VLT-SPHERE, and SCExAO will image dozens of gas giants, but the vast majority of known extrasolar planets have only been detected through indirect methods. The following are the indirect methods that have proven useful:

Indirect Methods

- Transit Method: If a planet crosses (or transits) in front of its parent star's disk, then the observed brightness of the star drops by a small amount. The amount by which the star dims depends on its size and on the size of the planet, among other factors. Because the transit method requires that the planet's orbit intersect a line-of-sight between the host star and Earth, the probability that an exoplanet in a randomly oriented orbit will be observed to transit the star is somewhat small. The Kepler telescope uses this method.
- Radial velocity or Doppler method: As a planet orbits a star, the star also moves in its own small orbit around the system's center of mass. Variations in the star's radial velocity–that is, the speed with which it moves towards or away from

Earth—can be detected from displacements in the star's spectral lines due to the Doppler effect. Extremely small radial-velocity variations can be observed, of 1 m/s or even somewhat less.

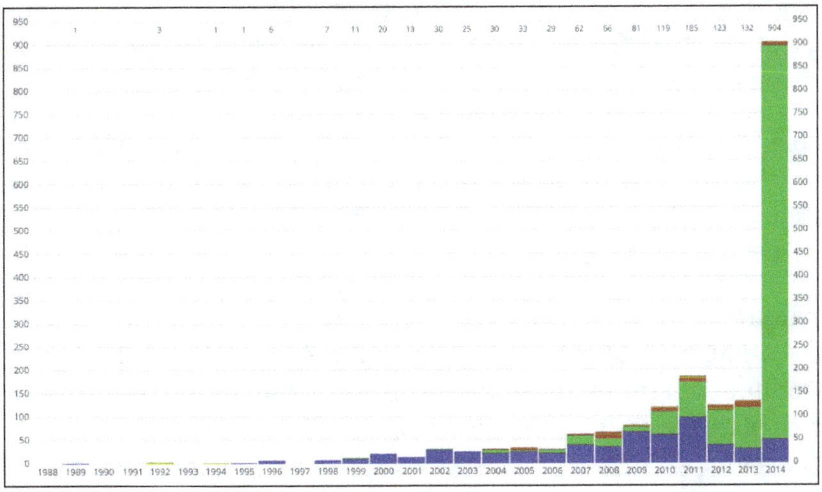

Discovered extrasolar planets per year and by detection method.
■ direct imaging ■ transit ■ radial velocity ■ microlensing ■ timing.

- Transit timing variation (TTV): When multiple planets are present, each one slightly perturbs the others' orbits. Small variations in the times of transit for one planet can thus indicate the presence of another planet, which itself may or may not transit. For example, variations in the transits of the planet Kepler-19b suggest the existence of a second planet in the system, the non-transiting Kepler-19c.

- Transit duration variation (TDV): When a planet orbits multiple stars or if the planet has moons, its transit time can significantly vary per transit. Although no new planets or moons have been discovered with this method, it is used to successfully confirm many transiting circumbinary planets.

- Gravitational microlensing: Microlensing occurs when the gravitational field of a star acts like a lens, magnifying the light of a distant background star. Planets orbiting the lensing star can cause detectable anomalies in the magnification as it varies over time. Unlike most other methods which have detection bias towards planets with small (or for resolved imaging, large) orbits, microlensing method is most sensitive to detecting planets around 1–10 AU away from Sun-like stars.

- Astrometry: Astrometry consists of precisely measuring a star's position in the sky and observing the changes in that position over time. The motion of a star due to the gravitational influence of a planet may be observable. Because the motion is so small, however, this method has not yet been very productive. It has produced only a few disputed detections, though it has been successfully used to investigate the properties of planets found in other ways.

- Pulsar timing: A pulsar (the small, ultradense remnant of a star that has exploded as a supernova) emits radio waves extremely regularly as it rotates. If planets orbit the pulsar, they will cause slight anomalies in the timing of its observed radio pulses. The first confirmed discovery of an extrasolar planet was made using this method. But as of 2011, it has not been very productive; five planets have been detected in this way, around three different pulsars.
- Variable star timing (pulsation frequency): Like pulsars, there are some other types of stars which exhibit periodic activity. Deviations from the periodicity can sometimes be caused by a planet orbiting it. As of 2013, a few planets have been discovered with this method.
- Reflection/emission modulations: When a planet orbits very close to the star, it catches a considerable amount of starlight. As the planet orbits around the star, the amount of light changes due to planets having phases from Earth's viewpoint or planet glowing more from one side than the other due to temperature differences.
- Relativistic beaming: Relativistic beaming measures the observed flux from the star due to its motion. The brightness of the star changes as the planet moves closer or further away from its host star.
- Ellipsoidal variations: Massive planets close to their host stars can slightly deform the shape of the star. This causes the brightness of the star to slightly deviate depending how it is rotated relative to Earth.
- Polarimetry: With polarimetry method, a polarized light reflected off the planet is separated from unpolarized light emitted from the star. No new planets have been discovered with this method although a few already discovered planets have been detected with this method.
- Circumstellar disks: Disks of space dust surround many stars, thought to originate from collisions among asteroids and comets. The dust can be detected because it absorbs starlight and re-emits it as infrared radiation. Features in the disks may suggest the presence of planets, though this is not considered a definitive detection method.

Orbital Parameters

Most known extrasolar planet candidates have been discovered using indirect methods and therefore only some of their physical and orbital parameters can be determined. For example, out of the six independent parameters that define an orbit, the radial-velocity method can determine four: semi-major axis, eccentricity, longitude of periastron, and time of periastron. Two parameters remain unknown: inclination and longitude of the ascending node.

Distance from Star and Orbital Period

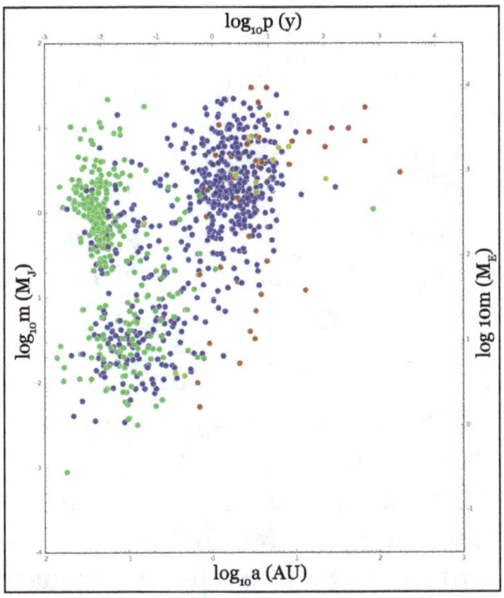

Log-log scatterplot showing masses, orbital radii, and period of all extrasolar planets discovered through September 2014, with colors indicating method of detection: transit radial velocity microlensing timing direct imaging.

There are exoplanets that are much closer to their parent star than any planet in the Solar System is to the Sun, and there are also exoplanets that are much further from their star. Mercury, the closest planet to the Sun at 0.4 astronomical units (AU), takes 88 days for an orbit, but the smallest known orbits of exoplanets have orbital periods of only a few hours, e.g. Kepler-70b. The Kepler-11 system has five of its planets in smaller orbits than Mercury's. Neptune is 30 AU from the Sun and takes 165 years to orbit it, but there are exoplanets that are thousands of AU from their star and take tens of thousands of years to orbit, e.g. GU Piscium b.

The radial-velocity and transit methods are most sensitive to planets with small orbits. The earliest discoveries such as 51 Peg b were gas giants with orbits of a few days. These "hot Jupiters" likely formed further out and migrated inwards. The *Kepler* space observatory has found planets with even shorter orbits of only a few hours, which places them within the star's upper atmosphere or corona, and these planets are Earth-sized or smaller and are probably the left-over solid cores of giant planets that have evaporated due to being so close to the star, or even being engulfed by the star in its red-giant phase in the case of Kepler-70b. As well as evaporation, other reasons why larger planets are unlikely to survive orbits only a few hours long include orbital decay caused by tidal force, tidal-inflation instability, and Roche-lobe overflow. The Roche limit implies that small planets with orbits of a few hours are likely made mostly of iron.

The direct imaging method is most sensitive to planets with large orbits, and has discovered some planets that have planet–star separations of hundreds of AU. However, protoplanetary disks are usually only around 100 AU in radius, and core accretion models predict giant planet formation to be within 10 AU, where the planets can coalesce quickly enough before the disk evaporates. Very-long-period giant planets may have been rogue planets that were captured, or formed close-in and gravitationally scattered outwards, or the planet and star could be a mass-imbalanced wide binary system with the planet being the primary object of its own separate protoplanetary disk. Gravitational instability models might produce planets at multi-hundred AU separations but this would require unusually large disks. For planets with very wide orbits up to several hundred thousand AU it may be difficult to observationally determine whether the planet is gravitationally bound to the star.

Most planets that have been discovered are within a couple of AU from their host star because the most used methods (radial-velocity and transit) require observation of several orbits to confirm that the planet exists and there has only been enough time since these methods were first used to cover small separations. Some planets with larger orbits have been discovered by direct imaging but there is a middle range of distances, roughly equivalent to the Solar System's gas giant region, which is largely unexplored. Direct imaging equipment for exploring that region was installed on two large telescopes that began operation in 2014, e.g. Gemini Planet Imager and VLT-SPHERE. The microlensing method has detected a few planets in the 1–10 AU range. It appears plausible that in most exoplanetary systems, there are one or two giant planets with orbits comparable in size to those of Jupiter and Saturn in the Solar System. Giant planets with substantially larger orbits are now known to be rare, at least around Sun-like stars.

The distance of the habitable zone from a star depends on the type of star and this distance changes during the star's lifetime as the size and temperature of the star changes.

Eccentricity

The eccentricity of an orbit is a measure of how elliptical (elongated) it is. All the planets of the Solar System except for Mercury have near-circular orbits ($e<0.1$). Most exoplanets with orbital periods of 20 days or less have near-circular orbits, i.e. very low eccentricity. That is thought to be due to tidal circularization: reduction of eccentricity over time due to gravitational interaction between two bodies. The mostly sub-Neptune-sized planets found by the *Kepler* spacecraft with short orbital periods have very circular orbits. By contrast, the giant planets with longer orbital periods discovered by radial-velocity methods have quite eccentric orbits. (As of July 2010, 55% of such exoplanets have eccentricities greater than 0.2, whereas 17% have eccentricities greater than 0.5.) Moderate to high eccentricities ($e>0.2$) of giant planets are not an observational selection effect, because a planet can be detected about equally well regardless of the eccentricity of its orbit. The statistical significance of elliptical orbits in the ensemble of observed giant planets is somewhat surprising, because current theories of planetary formation suggest that low-mass planets should have their orbital eccentricity circularized by gravitational interactions with the surrounding protoplanetary disk. However, as a planet grows more massive and its interaction with the disk becomes nonlinear, it may induce eccentric motion of the surrounding disk's gas, which in turn may excite the planet's orbital eccentricity. Low eccentricities are correlated with high multiplicity (number of planets in the system) based on the analysis of planets observed by the radial-velocity method.

For weak Doppler signals near the limits of the current detection ability, the eccentricity becomes poorly constrained and biased towards higher values. It is suggested that some of the high eccentricities reported for low-mass exoplanets may be overestimates, because simulations show that many observations are also consistent with two planets on circular orbits. Reported observations of single planets in moderately eccentric orbits have about a 15% chance of being a pair of planets. This misinterpretation is especially likely if the two planets orbit with a 2:1 resonance. With the exoplanet sample known in 2009, a group of astronomers estimated that "(1) around 35% of the published eccentric one-planet solutions are statistically indistinguishable from planetary systems in 2:1 orbital resonance, (2) another 40% cannot be statistically distinguished from a circular orbital solution" and "(3) planets with masses comparable to Earth could be hidden in known orbital solutions of eccentric super-Earths and Neptune mass planets".

Radial velocity surveys found exoplanet orbits beyond 0.1 AU to be eccentric, particularly for large planets. Transit data obtained by the *Kepler* spacecraft, is consistent with the RV surveys and also revealed that smaller planets tend to have less eccentric orbits.

Inclination vs. Spin–orbit Angle

Orbital inclination is the angle between a planet's orbital plane and another plane of reference. For exoplanets, the inclination is usually stated with respect to an observer on Earth: the angle used is that between the normal to the planet's orbital plane and the line of sight from Earth to the star. Therefore, most planets observed by the transit method are close to 90 degrees. Because the word 'inclination' is used in exoplanet studies for this line-of-sight inclination then the angle between the planet's orbit and the star's rotation must use a different word and is termed the spin–orbit angle or spin–orbit alignment. In most cases the orientation of the star's rotational axis is unknown. The Kepler spacecraft has found a few hundred multi-planet systems and in most of these systems the planets all orbit in nearly the same plane, much like the Solar System. However, a combination of astrometric and radial-velocity measurements has shown that some planetary systems contain planets whose orbital planes are significantly tilted relative to each other. More than half of hot Jupiters have orbital planes substantially misaligned with their parent star's rotation. A substantial fraction of hot-Jupiters even have retrograde orbits, meaning that they orbit in the opposite direction from the star's rotation. Rather than a planet's orbit having been disturbed, it may be that the star itself flipped early in their system's formation due to interactions between the star's magnetic field and the planet-forming disk.

Periastron Precession

Periastron precession is the rotation of a planet's orbit within the orbital plane, i.e. the axes of the ellipse change direction. In the Solar System, perturbations from other planets are the main cause, but for close-in exoplanets the largest factor can be tidal forces between the star and planet. For close-in exoplanets, the general relativistic contribution to the precession is also significant and can be orders of magnitude larger than the same effect for Mercury. Some exoplanets have significantly eccentric orbits, which makes it easier to detect the precession. The effect of general relativity can be detectable in timescales of about 10 years or less.

Nodal Precession

Nodal precession is rotation of a planet's orbital plane. Nodal precession is more easily seen as distinct from periastron precession when the orbital plane is inclined to the star's rotation, the extreme case being a polar orbit.

WASP-33 is a fast-rotating star that hosts a hot Jupiter in an almost polar orbit. The quadrupole mass moment and the proper angular momentum of the star are 1900 and 400 times, respectively, larger than those of the Sun. This causes significant classical and relativistic deviations from Kepler's laws. In particular, the fast rotation causes large nodal precession because of the star's oblateness and the Lense–Thirring effect.

Rotation and Axial Tilt

In April 2014, the first measurement of a planet's rotation period was announced: the length of day for the super-Jupiter gas giant Beta Pictoris b is 8 hours (based on the assumption that the axial tilt of the planet is small). With an equatorial rotational velocity of 25 km per second, this is faster than for the giant planets of the Solar System, in line with the expectation that the more massive a giant planet, the faster it spins. Beta Pictoris b's distance from its star is 9 AU. At such distances the rotation of Jovian planets is not slowed by tidal effects. Beta Pictoris b is still warm and young and over the next hundreds of millions of years, it will cool down and shrink to about the size of Jupiter, and if its angular momentum is preserved, then as it shrinks, the length of its day will decrease to about 3 hours and its equatorial rotation velocity will speed up to about 40 km/s. The images of Beta Pictoris b do not have high enough resolution to directly see details but doppler spectroscopy techniques were used to show that different parts of the planet were moving at different speeds and in opposite directions from which it was inferred that the planet is rotating. With the next generation of large ground-based telescopes it will be possible to use doppler imaging techniques to make a global map of the planet, like the mapping of the brown dwarf Luhman 16B in 2014.

Plot of equatorial spin velocity vs. mass for planets comparing Beta Pictoris b to the Solar System planets.

Origin of Spin and Tilt of Terrestrial Planets

Giant impacts have a large effect on the spin of terrestrial planets. The last few giant impacts during planetary formation tend to be the main determiner of a terrestrial planet's rotation rate. On average the spin angular velocity will be about 70% of the velocity that would cause the planet to break up and fly apart; the natural outcome of planetary embryo impacts at speeds slightly larger than escape velocity. In later stages terrestrial planet spin is also affected by impacts with planetesimals. During the giant impact stage, the thickness of a protoplanetary disk is far larger than the size of planetary embryos so collisions are equally likely to come from any direction in three-dimensions. This results in the axial tilt of accreted planets ranging from 0 to 180 degrees with any direction as likely as any other with both prograde and retrograde spins equally probable. Therefore, prograde spin with a small axial tilt, common for the Solar System's terrestrial planets except Venus, is not common in general for terrestrial planets built by giant impacts. The initial axial tilt of a planet determined by giant impacts can be substantially changed by stellar tides if the planet is close to its star and by satellite tides if the planet has a large satellite.

Tidal Effects

For most planets, the rotation period and axial tilt (also called obliquity) are not known, but a large number of planets have been detected with very short orbits (where tidal effects are greater) that will probably have reached an equilibrium rotation that can be predicted (i.e. tidal lock, spin–orbit resonances, and non-resonant equilibria such as retrograde rotation).

Gravitational tides tend to reduce the axial tilt to zero but over a longer timescale than the rotation rate reaches equilibrium. However, the presence of multiple planets in a system can cause axial tilt to be captured in a resonance called a Cassini state. There are small oscillations around this state and in the case of Mars these axial tilt variations are chaotic.

Hot Jupiters' close proximity to their host star means that their spin–orbit evolution is mostly due to the star's gravity and not the other effects. Hot Jupiters' rotation rate is not thought to be captured into spin–orbit resonance because of the way in which such a fluid-body reacts to tides; a planet like this therefore slows down into synchronous rotation if its orbit is circular, or, alternatively, it slows down into a non-synchronous rotation if its orbit is eccentric. Hot Jupiters are likely to evolve towards zero axial tilt even if they had been in a Cassini state during planetary migration when they were further from their star. Hot Jupiters' orbits will become more circular over time, however the presence of other planets in the system on eccentric orbits, even ones as small as Earth and as far away as the habitable zone, can continue to maintain the eccentricity of the Hot Jupiter so that the length of time for tidal circularization can be billions instead of millions of years.

The rotation rate of planet HD 80606 b is predicted to be about 1.9 days. HD 80606 b avoids spin–orbit resonance because it is a gas giant. The eccentricity of its orbit means that it avoids becoming tidally locked.

Physical Parameters

Mass

When a planet is found by the radial-velocity method, its orbital inclination i is unknown and can range from 0 to 90 degrees. The method is unable to determine the true mass (M) of the planet, but rather gives a lower limit for its mass, M sini. In a few cases an apparent exoplanet may be a more massive object such as a brown dwarf or red dwarf. However, the probability of a small value of i (say less than 30 degrees, which would give a true mass at least double the observed lower limit) is relatively low ($1-\sqrt{3}/2 \approx 13\%$) and hence most planets will have true masses fairly close to the observed lower limit.

If a planet's orbit is nearly perpendicular to the line of vision (i.e. i close to 90°), a planet can be detected through the transit method. The inclination will then be known, and the inclination combined with M sini from radial-velocity observations will give the planet's true mass.

Also, astrometric observations and dynamical considerations in multiple-planet systems can sometimes provide an upper limit to the planet's true mass.

The mass of a transiting exoplanet can also be determined from the transmission spectrum of its atmosphere, as it can be used to constrain independently the atmospheric composition, temperature, pressure, and scale height.

Transit-timing variation can also be used to find a planet's mass.

Radius, Density and Bulk Composition

Prior to recent results from the Kepler space observatory, most confirmed planets were gas giants comparable in size to Jupiter or larger because they are most easily detected. However, the planets detected by Kepler are mostly between the size of Neptune and the size of Earth.

If a planet is detectable by both the radial-velocity and the transit methods, then both its true mass and its radius can be determined, as well as its density. Planets with low density are inferred to be composed mainly of hydrogen and helium, whereas planets of intermediate density are inferred to have water as a major constituent. A planet of high density is inferred to be rocky, like Earth and the other terrestrial planets of the Solar System.

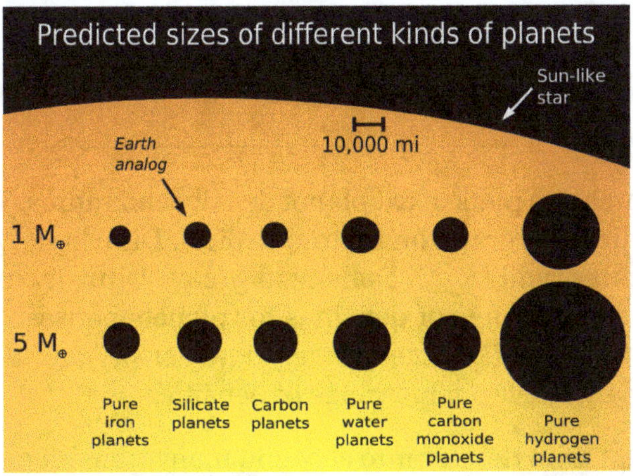

Comparison of sizes of planets with different compositions.

Gas Giants, Puffy Planets and Super-Jupiters

Size comparison of WASP-17b (right) with Jupiter (left).

Gaseous planets that are hot is because they are close to their star, or because they are still hot from their formation and are expanded by the heat. For colder gas planets, there is a maximum radius which is slightly larger than Jupiter which occurs when the mass reaches a few Jupiter-masses. Adding mass beyond this point causes the radius to shrink.

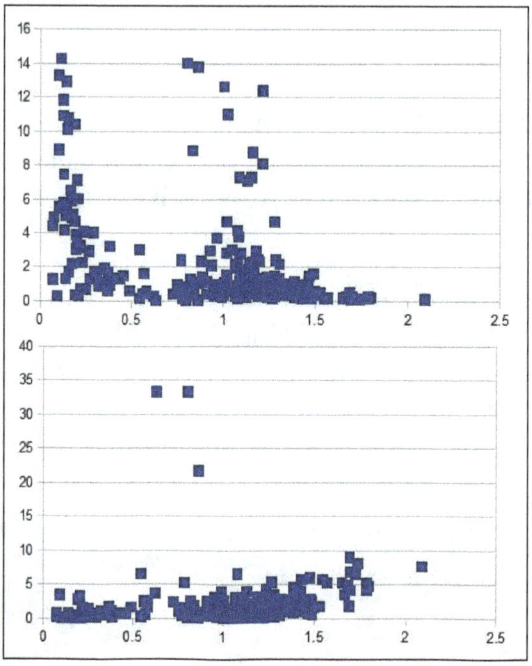

The image above represents plots of exoplanet density and radius. The graph on top depicts Density vs. Radius. the graph on bottom depicts, Diffusity= 1/Density vs. Radius. Units: Radius in Jupiter radii (R_{Jup}). Density in g/cm^3. Diffusity in cm^3/g. These plots show that there are a wide range of densities for planets between Earth and Neptune size, then the planets of 0.6 R_{Jup} size are very low-density and there are very few of them, then the gas giants have a large range of densities.

Even when taking heat from the star into account, many transiting exoplanets are much larger than expected given their mass, meaning that they have surprisingly low density.

Besides the inflated hot Jupiters, there is another type of low-density planet: occurring at around 0.6 times the size of Jupiter. The planets around Kepler-51 are far less dense (far more diffuse) than the inflated hot Jupiters as can be seen in the plots on the right where the three Kepler-51 planets stand out in the diffusity vs. radius plot. A more detailed study taking into account star spots may modify these results to produce less extreme values.

Ice Giants and Super-Neptunes

Kepler-101b was the first super-Neptune discovered. It has three times Neptune's mass but its density suggests that heavy elements make up more than 60% of its total mass, unlike hydrogen–helium-dominated gas giants.

Super-Earths, Mini-Neptunes and Gas Dwarfs

If a planet has a radius and mass between that of Earth and Neptune, then there is a question about whether the planet is rocky like Earth, a mixture of volatiles and gas like Neptune, a small planet with a hydrogen/helium envelope (mini-Jupiter), or of some other composition.

Some of the Kepler transiting planets with radii in the range of 1–4 Earth radii have had their masses measured by radial-velocity or transit-timing methods. The calculated densities show that up to 1.5 Earth radii, these planets are rocky and that density increases with increasing radius due to gravitational compression. However, between 1.5 and 4 Earth radii the density decreases with increasing radius. This indicates that above 1.5 Earth radii, planets tend to have increasing amounts of volatiles and gas. Despite this general trend, there is a wide range of masses at a given radius, which could be because gas planets can have rocky cores of different masses and compositions, and could also be due to photoevaporation of volatiles. Thermal evolutionary atmosphere models suggest a radius of 1.75 times that of Earth as a dividing line between rocky and gaseous planets. Excluding close-in planets that have lost their gas envelope due to stellar irradiation, studies of the metallicity of stars suggest a dividing line of 1.7 Earth radii between rocky planets and gas dwarfs, then another dividing line at 3.9 Earth radii between gas dwarfs and gas giants. These dividing lines are statistical trends and do not apply universally, because there are many other factors besides metallicity that affect planet formation, including distance from star – there may be larger rocky planets that formed at larger distances. An independent reanalysis of the data suggests that there are no such dividing lines and that there is a continuum of planet formation between 1 and 4 Earth radii and no reason to suspect that the amount of solid material in a protoplanetary disk determines whether super-Earths or mini-Neptunes form. Studies done in 2016 based on over 300 planets suggest that most objects over approximately two Earth masses collect significant hydrogen–helium envelopes, meaning rocky super-Earths may be rare.

The discovery of the low-density Earth-mass planet Kepler-138d shows that there is an overlapping range of masses in which both rocky planets and low-density planets occur. A low-mass low-density planets could be an ocean planet or super-Earth with a remnant hydrogen atmosphere, or a hot planet with a steam atmosphere, or a mini-Neptune with a hydrogen–helium atmosphere. Another possibility for a low-mass low-density planet is that it has a large atmosphere made up chiefly of carbon monoxide, carbon dioxide, methane, or nitrogen.

Massive Solid Planets

In 2014, new measurements of Kepler-10c found it to be a Neptune-mass planet (17 Earth masses) with a density higher than Earth's, indicating that Kepler-10c is composed mostly of rock with possibly up to 20% high-pressure water ice but without a hydrogen-dominated envelope. Because this is well above the 10-Earth-mass upper limit that is commonly used for the term 'super-Earth', the term mega-Earth has been coined. A similarly massive and dense planet could be Kepler-131b, although its density is not as well measured as that of Kepler 10c. The next most massive known solid planets are half this mass: 55 Cancri e and Kepler-20b.

Size comparison of Kepler-10c with Earth and Neptune.

Gas planets can have large solid cores. The Saturn-mass planet HD 149026 b has only two-thirds of Saturn's radius, so it may have a rock–ice core of 60 Earth masses or more. Corot-20b has 4.24 times Jupiter's mass but a radius of only 0.84 that of Jupiter; it may have a metal core of 800 Earth masses if the heavy elements are concentrated in the core, or a core of 300 Earth masses if the heavy elements are more distributed throughout the planet.

Transit-timing variation measurements indicate that Kepler-52b, Kepler-52c and Kepler-57b have maximum masses between 30 and 100 times that of Earth, although the actual masses could be much lower. With radii about 2 Earth radii in size, they might have densities larger than that of an iron planet of the same size. They orbit very close to their stars, so they could each be the remnant core (chthonian planet) of an evaporated gas giant or brown dwarf. If a remnant core is massive enough it could remain in such a state for billions of years despite having lost the atmospheric mass.

Solid planets up to thousands of Earth masses may be able to form around massive stars (B-type and O-type stars; 5–120 solar masses), where the protoplanetary disk would contain enough heavy elements. Also, these stars have high UV radiation and winds that could photoevaporate the gas in the disk, leaving just the heavy elements. For comparison, Neptune's mass equals 17 Earth masses, Jupiter has 318 Earth masses, and the 13-Jupiter-mass limit used in the IAU's working definition of an exoplanet equals approximately 4000 Earth masses.

Cold planets have a maximum radius because adding more mass at that point causes the planet to compress under the weight instead of increasing the radius. The maximum radius for solid planets is lower than the maximum radius for gas planets.

Shape

When the size of a planet is described using its radius, this is approximating the shape by a sphere. However, the rotation of a planet causes it to be flattened at the poles; so the equatorial radius is larger than the polar radius, making it closer to an oblate spheroid. The oblateness of transiting exoplanets will affect the transit light curves. At the limits of current technology it has been possible to show that HD 189733b is less oblate than Saturn. If the planet is close to its star, then gravitational tides will elongate the planet in the direction of the star, making the planet closer to a triaxial ellipsoid. Because tidal deformation is along a line between the planet and the star, it is difficult to detect from transit photometry; it will have an effect on the transit light curves an order of magnitude less than that caused by rotational deformation even in cases where tidal deformation is larger than rotational deformation (as is the case for tidally locked hot Jupiters). Material rigidity of rocky planets and rocky cores of gas planets will cause further deviations from the aforementioned shapes. Thermal tides caused by unevenly irradiated surfaces are another factor.

VISIBLE-LIGHT ASTRONOMY

Visible-light astronomy encompasses a wide variety of observations via telescopes that are sensitive in the range visible light (optical telescopes). Visible-light astronomy is part of optical astronomy, and differs from astronomies based on invisible types of light in the electromagnetic radiation spectrum, such as radio waves, infrared waves, ultraviolet waves, X-ray waves and gamma-ray waves. Visible light ranges from 380 to 750 nanometers in wavelength.

Visible-light astronomy has existed as long as people have been looking up at the night sky, although it has since improved in its observational capabilities since the invention of the telescope, which is commonly credited to Hans Lippershey, a German-Dutch spectacle-maker, although Galileo played a large role in the development and creation of telescopes. Visible-light astronomy continues to get better in the modern day, with projects such as the James Webb Telescope being projected for launch in the next few years.

Since visible-light astronomy is restricted to only visible light, no equipment is necessary for simply star gazing. This means that it's the most commonly participated in type of astronomy, as well as the oldest.

Effect of Ambient Brightness

The visibility of celestial objects in the night sky is affected by light pollution, with the presence of the Moon in the night sky historically hindering astronomical observation by increasing the amount of ambient lighting. With the advent of artificial light sources, however, light pollution has been a growing problem for viewing the night sky. Special filters and modifications to light fixtures can help to alleviate this problem, but for the best views, both professional and amateur optical astronomers seek viewing sites located far from major urban areas. In order to avoid light pollution of Earth's sky, among other reasons, many telescopes are put outside of the Earth's atmosphere, where not only light pollution, but also atmospheric distortion and obscuration are minimized.

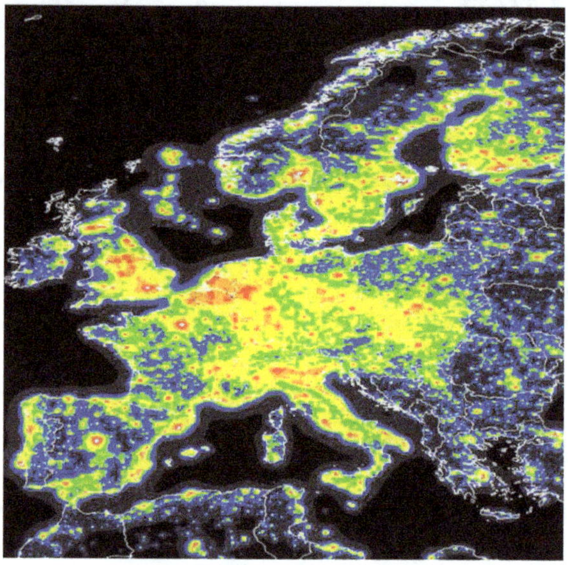

Light pollution map of Europe.

Commonly Observed Objects

The most commonly observed objects tend to be ones that do not require a telescope to view, such as the Moon, meteors, planets, constellations, and stars.

The Moon is a very commonly observed astronomical object, especially by amateur astronomers and skygazers. This is due to several reasons: the Moon is the brightest object in the night sky, the Moon is the largest object in the night sky, and the Moon has long been significant in many cultures, such as being the basis for many calendars. The Moon also does not require any kind of telescope or binoculars to see effectively, making it extremely convenient and common for people to observe.

Meteors, often called "shooting stars" are also commonly observed. Meteor showers, such as the Perseids and Leonids, make viewing meteors much easier, as a multitude of meteors are visible in a relatively short period of time.

Planets are usually observed with the aid of a telescope or binoculars. Venus is likely the easiest planet to observe without the aid of any instruments, as it is very bright, and can even be seen in daylight. However, Mars, Jupiter, and Saturn can also be seen without the aid of telescopes or binoculars.

Constellations and stars are also often observed, and have been used in the past for navigation, especially by ships at sea. One of the most recognizable constellations is the Big Dipper, which is part of the constellation Ursa Major. Constellations also serve to help describe the location of other objects in the sky.

ULTRAVIOLET ASTRONOMY

Ultraviolet astronomy is the study of the ultraviolet spectra of astronomical objects. Ultraviolet radiation comes from a hotter region of the electromagnetic spectrum than visible light. For example, interstellar gas at temperatures close to 1,000,000 kelvins is quite prominent in the ultraviolet. It has yielded much important information about chemical abundances and processes in the Sun and certain other stellar objects, such as white dwarfs.

Ultraviolet astronomy became feasible with the advent of rockets capable of carrying instruments above Earth's atmosphere, which absorbs most electromagnetic radiation of ultraviolet wavelengths (i.e., roughly 100 to 4,000 angstroms) from celestial sources. Much radiation is lost even at the highest altitudes that balloons can reach. During the 1920s, unsuccessful attempts were made to photograph the Sun's ultraviolet spectrum from balloons; not until 1946 did a rocket-borne camera succeed in doing so.

Since the early 1960s the United States and several other countries have placed in Earth orbit unmanned satellite observatories carrying telescopes with optical surfaces specially coated for high ultraviolet reflectivity. These include eight Orbiting Solar Observatories, launched from 1962 to 1975 by the U.S. National Aeronautics and Space Administration (NASA), which enabled astronomers to obtain thousands of ultraviolet spectra of the Sun's corona. Another series of U.S. satellites, known as Orbiting Astronomical Observatories, in service from 1968 to 1981, permitted the study of the interstellar medium and remote stars in the spectral range of 1,200 to 4,000 angstroms.

A telescope carried aboard the International Ultraviolet Explorer spacecraft (launched in 1978 by the European Space Agency [ESA], NASA, and the United Kingdom) allowed significant ultraviolet observations to be made of objects such as comets and quasars. The high-resolution Hubble Space Telescope, deployed in 1990, also collected ultraviolet-wavelength data about faint objects such as nebulae and distant star clusters. NASA's Extreme Ultraviolet Explorer (EUVE) satellite was launched in 1992 and studied stellar evolution and the interstellar medium. EUVE was succeeded in 1999 by NASA's Far Ultraviolet Spectroscopic Explorer (FUSE), which discovered molecular nitrogen in interstellar space. Another NASA ultraviolet satellite, the Galaxy Evolution Explorer (GALEX), was launched in 2003 and studied how galaxies change over billions of years. The Solar and Heliospheric Observatory (SOHO), an ESA-NASA satellite launched in 1995, has studied the Sun and its hot corona in ultraviolet light.

Ultraviolet–visible Spectroscopy

Ultraviolet–visible spectroscopy or ultraviolet–visible spectrophotometry (UV–Vis or UV/Vis) refers to absorption spectroscopy or reflectance spectroscopy in part of the ultraviolet and the full, adjacent visible spectral regions. This means it uses light in the visible and adjacent ranges. The absorption or reflectance in the visible range directly affects the perceived color of the chemicals involved. In this region of the electromagnetic spectrum, atoms and molecules undergo electronic transitions. Absorption spectroscopy is complementary to fluorescence spectroscopy, in that fluorescence deals with transitions from the excited state to the ground state, while absorption measures transitions from the ground state to the excited state.

Principle of Ultraviolet–visible Absorption

Molecules containing bonding and non-bonding electrons (n-electrons) can absorb energy in the form of ultraviolet or visible light to excite these electrons to higher anti-bonding molecular orbitals. The more easily excited the electrons (i.e. lower energy gap between the HOMO and the LUMO), the longer the wavelength of light it can absorb. There are four possible types of transitions (π–π*, n–π*, σ–σ*, and n–σ*), and they can be ordered as follows: σ–σ* > n–σ* > π–π* > n–π*.

Applications

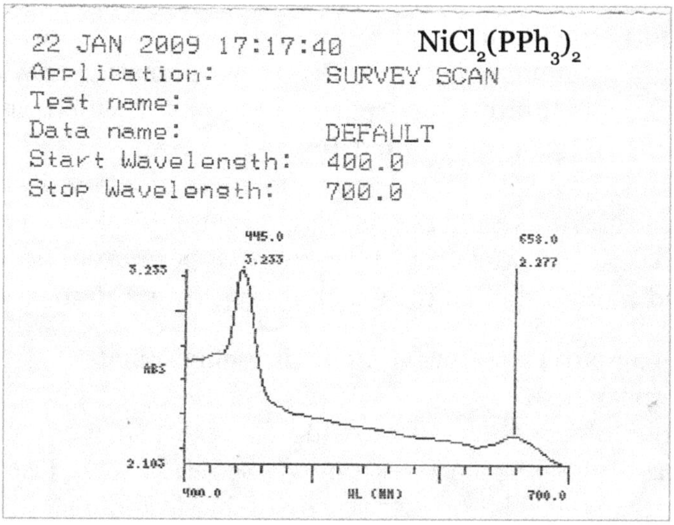

An example of a UV/Vis readout.

UV/Vis spectroscopy is routinely used in analytical chemistry for the quantitative determination of different analytes, such as transition metal ions, highly conjugated organic compounds, and biological macromolecules. Spectroscopic analysis is commonly carried out in solutions but solids and gases may also be studied.

- Solutions of transition metal ions can be colored (i.e., absorb visible light) because d electrons within the metal atoms can be excited from one electronic state to another. The colour of metal ion solutions is strongly affected by the presence of other species, such as certain anions or ligands. For instance, the colour of a dilute solution of copper sulfate is a very light blue; adding ammonia intensifies the colour and changes the wavelength of maximum absorption (λ_{max}).

- Organic compounds, especially those with a high degree of conjugation, also absorb light in the UV or visible regions of the electromagnetic spectrum. The solvents for these determinations are often water for water-soluble compounds, or ethanol for organic-soluble compounds. (Organic solvents may have significant UV absorption; not all solvents are suitable for use in UV spectroscopy. Ethanol absorbs very weakly at most wavelengths.) Solvent polarity and pH can affect the absorption spectrum of an organic compound. Tyrosine, for example, increases in absorption maxima and molar extinction coefficient when pH increases from 6 to 13 or when solvent polarity decreases.
- While charge transfer complexes also give rise to colours, the colours are often too intense to be used for quantitative measurement.

The Beer–Lambert law states that the absorbance of a solution is directly proportional to the concentration of the absorbing species in the solution and the path length. Thus, for a fixed path length, UV/Vis spectroscopy can be used to determine the concentration of the absorber in a solution. It is necessary to know how quickly the absorbance changes with concentration. This can be taken from references (tables of molar extinction coefficients), or more accurately, determined from a calibration curve.

A UV/Vis spectrophotometer may be used as a detector for HPLC. The presence of an analyte gives a response assumed to be proportional to the concentration. For accurate results, the instrument's response to the analyte in the unknown should be compared with the response to a standard; this is very similar to the use of calibration curves. The response (e.g., peak height) for a particular concentration is known as the response factor.

The wavelengths of absorption peaks can be correlated with the types of bonds in a given molecule and are valuable in determining the functional groups within a molecule. The Woodward–Fieser rules, for instance, are a set of empirical observations used to predict λ_{max}, the wavelength of the most intense UV/Vis absorption, for conjugated organic compounds such as dienes and ketones. The spectrum alone is not, however, a specific test for any given sample. The nature of the solvent, the pH of the solution, temperature, high electrolyte concentrations, and the presence of interfering substances can influence the absorption spectrum. Experimental variations such as the slit width (effective bandwidth) of the spectrophotometer will also alter the spectrum. To apply UV/Vis spectroscopy to analysis, these variables must be controlled or accounted for in order to identify the substances present.

The method is most often used in a quantitative way to determine concentrations of an absorbing species in solution, using the Beer–Lambert law:

$$A = \log_{10}(I_0 / I) = \varepsilon c L ,$$

where A is the measured absorbance (in Absorbance Units (AU)), I_0 is the intensity of the incident light at a given wavelength, I is the transmitted intensity, L the path length through the sample, and c the concentration of the absorbing species. For each species and wavelength, ε is a constant known as the molar absorptivity or extinction coefficient. This constant is a fundamental molecular property in a given solvent, at a particular temperature and pressure, and has units of $1 / M * cm$.

The absorbance and extinction ε are sometimes defined in terms of the natural logarithm instead of the base-10 logarithm.

The Beer–Lambert Law is useful for characterizing many compounds but does not hold as a universal relationship for the concentration and absorption of all substances. A 2nd order polynomial relationship between absorption and concentration is sometimes encountered for very large, complex molecules such as organic dyes (Xylenol Orange or Neutral Red, for example).

UV–Vis spectroscopy is also used in the semiconductor industry to measure the thickness and optical properties of thin films on a wafer. UV–Vis spectrometers are used to measure the reflectance of light, and can be analyzed via the Forouhi–Bloomer dispersion equations to determine the Index of Refraction (n) and the Extinction Coefficient (k) of a given film across the measured spectral range.

Practical Considerations

The Beer–Lambert law has implicit assumptions that must be met experimentally for it to apply; otherwise there is a possibility of deviations from the law. For instance, the chemical makeup and physical environment of the sample can alter its extinction coefficient. The chemical and physical conditions of a test sample therefore must match reference measurements for conclusions to be valid. Worldwide, pharmacopoeias such as the American (USP) and European (Ph. Eur.) pharmacopeias demand that spectrophotometers perform according to strict regulatory requirements encompassing factors such as stray light and wavelength accuracy.

Spectral Bandwidth

It is important to have a monochromatic source of radiation for the light incident on the sample cell. Monochromaticity is measured as the width of the "triangle" formed by the intensity spike, at one half of the peak intensity. A given spectrometer has a spectral bandwidth that characterizes how monochromatic the incident light is. If this bandwidth is comparable to (or more than) the width of the absorption line, then the measured extinction coefficient will be mistaken. In reference measurements, the instrument bandwidth (bandwidth of the incident light) is kept below the width of the spectral lines. When a test material is being measured, the bandwidth of the incident light should also be sufficiently narrow. Reducing the spectral bandwidth reduces the energy passed to the detector and will, therefore, require a longer measurement time to achieve the same signal to noise ratio.

Wavelength Error

In liquids, the extinction coefficient usually changes slowly with wavelength. A peak of the absorbance curve (a wavelength where the absorbance reaches a maximum) is where the rate of change in absorbance with wavelength is smallest. Measurements are usually made at a peak to minimize errors produced by errors in wavelength in the instrument, that is errors due to having a different extinction coefficient than assumed.

Stray Light

Another important factor is the purity of the light used. The most important factor affecting this is the stray light level of the monochromator.

The detector used is broadband; it responds to all the light that reaches it. If a significant amount of the light passed through the sample contains wavelengths that have much lower extinction coefficients than the nominal one, the instrument will report an incorrectly low absorbance. Any instrument will reach a point where an increase in sample concentration will not result in an increase in the reported absorbance, because the detector is simply responding to the stray light. In practice the concentration of the sample or the optical path length must be adjusted to place the unknown absorbance within a range that is valid for the instrument. Sometimes an empirical calibration function is developed, using known concentrations of the sample, to allow measurements into the region where the instrument is becoming non-linear.

As a rough guide, an instrument with a single monochromator would typically have a stray light level corresponding to about 3 Absorbance Units (AU), which would make measurements above about 2 AU problematic. A more complex instrument with a double monochromator would have a stray light level corresponding to about 6 AU, which would therefore allow measuring a much wider absorbance range.

Deviations from the Beer–Lambert Law

At sufficiently high concentrations, the absorption bands will saturate and show absorption flattening. The absorption peak appears to flatten because close to 100% of the light is already being absorbed. The concentration at which this occurs depends on the particular compound being measured. One test that can be used to test for this effect is to vary the path length of the measurement. In the Beer–Lambert law, varying concentration and path length has an equivalent effect—diluting a solution by a factor of 10 has the same effect as shortening the path length by a factor of 10. If cells of different path lengths are available, testing if this relationship holds true is one way to judge if absorption flattening is occurring.

Solutions that are not homogeneous can show deviations from the Beer–Lambert law because of the phenomenon of absorption flattening. This can happen, for instance, where the absorbing substance is located within suspended particles. The deviations will be most noticeable under conditions of low concentration and high absorbance. The last reference describes a way to correct for this deviation.

Some solutions, like copper(II)chloride in water, change visually at a certain concentration because of changed conditions around the coloured ion (the divalent copper ion). For copper(II)chloride it means a shift from blue to green, which would mean that monochromatic measurements would deviate from the Beer–Lambert law.

Measurement Uncertainty Sources

The above factors contribute to the measurement uncertainty of the results obtained with UV/Vis spectrophotometry. If UV/Vis spectrophotometry is used in quantitative chemical analysis then the results are additionally affected by uncertainty sources arising from the nature of the compounds and/or solutions that are measured. These include spectral interferences caused by absorption band overlap, fading of the color of the absorbing species (caused by decomposition or reaction) and possible composition mismatch between the sample and the calibration solution.

Ultraviolet–visible Spectrophotometer

The instrument used in ultraviolet–visible spectroscopy is called a UV/Vis spectrophotometer. It measures the intensity of light passing through a sample (I), and compares it to the intensity of light before it passes through the sample (I_o). The ratio I / I_o is called the transmittance, and is usually expressed as a percentage (%T). The absorbance, A, is based on the transmittance:

$$A = -\log(\%T / 100\%)$$

The UV–visible spectrophotometer can also be configured to measure reflectance. In this case, the spectrophotometer measures the intensity of light reflected from a sample (I), and compares it to the intensity of light reflected from a reference material (I_o) (such as a white tile). The ratio I / I_o is called the reflectance, and is usually expressed as a percentage (%R).

The basic parts of a spectrophotometer are a light source, a holder for the sample, a diffraction grating in a monochromator or a prism to separate the different wavelengths of light, and a detector. The radiation source is often a Tungsten filament (300–2500 nm), a deuterium arc lamp, which is continuous over the ultraviolet region (190–400 nm), Xenon arc lamp, which is continuous from 160 to 2,000 nm; or more recently, light emitting diodes (LED) for the visible wavelengths. The detector is typically a photomultiplier tube, a photodiode, a photodiode array or a charge-coupled device (CCD). Single photodiode detectors and photomultiplier tubes are used with scanning monochromators, which filter the light so that only light of a single wavelength reaches the detector at one time. The scanning monochromator moves the diffraction grating to "step-through" each wavelength so that its intensity may be measured as a function of wavelength. Fixed monochromators are used with CCDs and photodiode arrays. As both of these devices consist of many detectors grouped into one or two dimensional arrays, they are able to collect light of different wavelengths on different pixels or groups of pixels simultaneously.

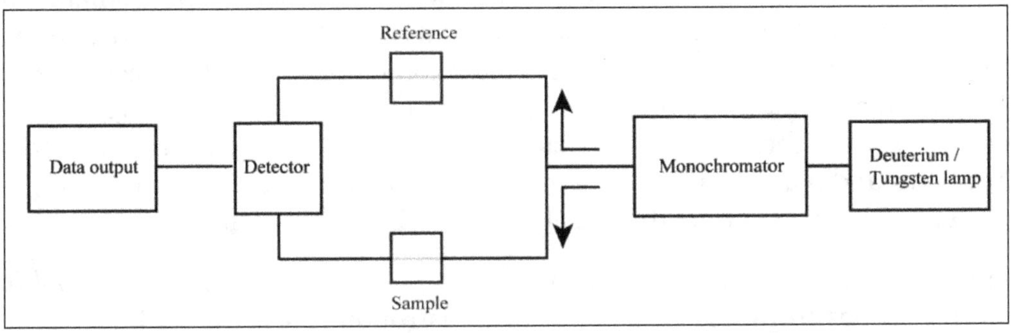

Simplified schematic of a double beam UV–visible spectrophotometer.

A spectrophotometer can be either single beam or double beam. In a single beam instrument (such as the Spectronic 20), all of the light passes through the sample cell. must be measured by removing the sample. This was the earliest design and is still in common use in both teaching and industrial labs.

In a double-beam instrument, the light is split into two beams before it reaches the sample. One beam is used as the reference; the other beam passes through the sample. The reference beam intensity is taken as 100% Transmission (or 0 Absorbance), and the measurement displayed is the ratio of the two beam intensities. Some double-beam instruments have two detectors (photodiodes), and the sample and reference beam are measured at the same time. In other instruments, the two beams pass through a beam chopper, which blocks one beam at a time. The detector alternates between measuring the sample beam and the reference beam in synchronism with the chopper. There may also be one or more dark intervals in the chopper cycle. In this case, the measured beam intensities may be corrected by subtracting the intensity measured in the dark interval before the ratio is taken.

In a single-beam instrument, the cuvette containing only a solvent has to be measured first. Mettler Toledo developed a single beam array spectrophotometer that allows fast and accurate measurements over the UV/VIS range. The light source consists of a

Xenon flash lamp for the ultraviolet (UV) as well as for the visible (VIS) and near-infrared wavelength regions covering a spectral range from 190 up to 1100 nm. The lamp flashes are focused on a glass fiber which drives the beam of light onto a cuvette containing the sample solution. The beam passes through the sample and specific wavelengths are absorbed by the sample components. The remaining light is collected after the cuvette by a glass fiber and driven into a spectrograph. The spectrograph consists of a diffraction grating that separates the light into the different wavelengths, and a CCD sensor to record the data, respectively. The whole spectrum is thus simultaneously measured, allowing for fast recording.

Samples for UV/Vis spectrophotometry are most often liquids, although the absorbance of gases and even of solids can also be measured. Samples are typically placed in a transparent cell, known as a cuvette. Cuvettes are typically rectangular in shape, commonly with an internal width of 1 cm. (This width becomes the path length, , in the Beer–Lambert law.) Test tubes can also be used as cuvettes in some instruments. The type of sample container used must allow radiation to pass over the spectral region of interest. The most widely applicable cuvettes are made of high quality fused silica or quartz glass because these are transparent throughout the UV, visible and near infrared regions. Glass and plastic cuvettes are also common, although glass and most plastics absorb in the UV, which limits their usefulness to visible wavelengths.

Specialized instruments have also been made. These include attaching spectrophotometers to telescopes to measure the spectra of astronomical features. UV–visible microspectrophotometers consist of a UV–visible microscope integrated with a UV–visible spectrophotometer.

A complete spectrum of the absorption at all wavelengths of interest can often be produced directly by a more sophisticated spectrophotometer. In simpler instruments the absorption is determined one wavelength at a time and then compiled into a spectrum by the operator. By removing the concentration dependence, the extinction coefficient (ε) can be determined as a function of wavelength.

Microspectrophotometry

UV–visible spectroscopy of microscopic samples is done by integrating an optical microscope with UV–visible optics, white light sources, a monochromator, and a sensitive detector such as a charge-coupled device (CCD) or photomultiplier tube (PMT). As only a single optical path is available, these are single beam instruments. Modern instruments are capable of measuring UV–visible spectra in both reflectance and transmission of micron-scale sampling areas. The advantages of using such instruments is that they are able to measure microscopic samples but are also able to measure the spectra of larger samples with high spatial resolution. As such, they are used in the forensic laboratory to analyze the dyes and pigments in individual textile fibers, microscopic paint chips and the color of glass fragments. They are also used in materials science and biological research and for determining the energy content of coal and petroleum source rock by measuring the vitrinite reflectance. Microspectrophotometers are used in the semiconductor and micro-optics industries for monitoring the thickness of thin films after they have been deposited. In the semiconductor industry, they are used because the critical dimensions of circuitry is microscopic. A typical test of a semiconductor wafer would entail the acquisition of spectra from many points on a patterned or unpatterned wafer. The thickness of the deposited films may be calculated from the interference pattern of the spectra. In addition, ultraviolet–visible spectrophotometry can be used to determine the thickness, along with the refractive index and extinction coefficient of thin films as described in Refractive index and extinction coefficient of thin film materials. A map of the film thickness across the entire wafer can then be generated and used for quality control purposes.

Additional Applications

UV/Vis can be applied to determine the kinetics or rate constant of a chemical reaction. The reaction, occurring in solution, must present color or brightness shifts from reactants to products in order to use UV/Vis for this application. For example, the molecule mercury dithizonate is a yellow-orange color in diluted solution (1*10^-5 M), and turns blue when subjected with particular wavelengths of visible light (and UV) via a conformational change, but this reaction is reversible back into the yellow "ground state".

The rate constant of a particular reaction can be determined by measuring the UV/Vis absorbance spectrum at specific time intervals. Using mercury dithizonate again as an example, one can shine light on the sample to turn the solution blue, then run a UV/Vis test every 10 seconds (variable) to see the levels of absorbed and reflected wavelengths change over time in accordance with the solution turning back to yellow from the excited blue energy state. From these measurements, the concentration of the two species can be calculated. The mercury dithizonate reaction from one conformation to another is first order and would have the integral first order rate law : ln[A](time t)=−kt+ln[A] (initial). Therefore, graphing the natural log (ln) of the concentration [A] versus time will graph a line with slope -k, or negative the rate constant. Different rate orders have different integrated rate laws depending on the mechanism of the reaction.

An equilibrium constant can also be calculated with UV/Vis spectroscopy. After determining optimal wavelengths for all species involved in equilibria, a reaction can be run to equilibrium, and the concentration of species determined from spectroscopy at various known wavelengths. The equilibrium constant can be calculated as K(eq) = [Products] / [Reactants].

X-RAY ASTRONOMY

X-ray astronomy is an observational branch of astronomy that focuses on the study of celestial objects based on their X-ray emissions. These emissions are thought to come from sources that contain extremely hot matter, at temperatures ranging from a million to hundred million kelvin (K). This matter is in a state known as plasma (ionized gas), which consists of ions and electrons at very high energies.

Astronomers have discovered various types of X-ray sources in the universe. They include stars, binary stars containing a white dwarf, neutron stars, supernova remnants, galaxy clusters, and black holes. Some Solar System bodies, such as the Moon, also emit X-rays, although most of the X-ray brightness of the Moon arises from reflected solar X-rays. The detection of X-rays gives scientists clues about possible processes and events that may be occurring at or near the radiation sources.

This image of the Crab Nebula was created by superimposing the X-ray (blue), and optical (red) images. The X-ray image is smaller because the higher energy X-ray emitting electrons radiate away their energy faster than the lower energy optically emitting electrons as they move.

How Astronomers Observe X-rays

Nearly all of the X-ray radiation from cosmic sources is absorbed by the Earth's atmosphere. X-rays that have energies in the 0.5 to 5 keV (80 to 800 aJ) range, in which most celestial sources give off the bulk of their energy, can be stopped by a few sheets of paper. Ninety percent of the photons in a beam of three keV (480 aJ) X-rays are absorbed by traveling through just ten cm of air. Even highly energetic X-rays, consisting of photons at energies greater than 30 keV (4,800 aJ), can penetrate through only a few meters of the atmosphere.

For this reason, to observe X-rays from the sky, the detectors must be flown above most of the Earth's atmosphere. In the past, X-ray detectors were carried by balloons and sounding rockets. Nowadays, scientists prefer to put the detectors on satellites.

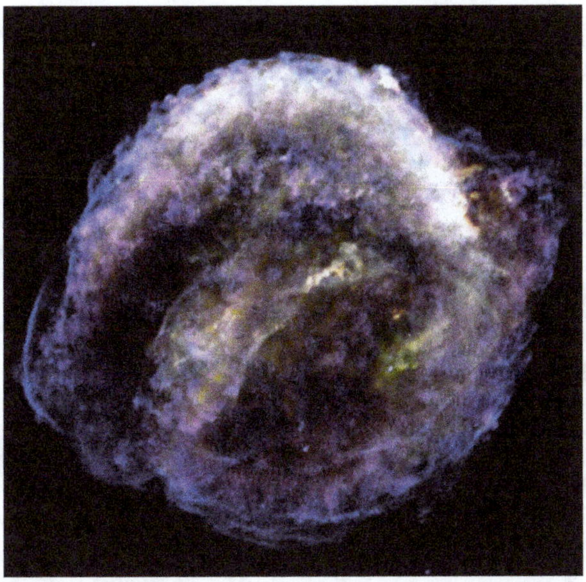

This X-ray image, reveals remnants of Kepler's Supernova.

Sounding Rocket Flights

An X-ray detector may be placed in the nose cone section of a sounding rocket and launched above the atmosphere. This was first done at White Sands Missile Range in New Mexico with a V-2 rocket in 1949. X-rays from the Sun were detected by the Navy's experiment on board. In June 1962, an instrument aboard an Aerobee 150 rocket first detected X-rays from another celestial source.

The greatest drawbacks to rocket flights are (a) their very short duration (just a few minutes above the atmosphere before the rocket falls back to Earth), and (b) their limited field of view. A rocket launched from the United States will not be able to see sources in the southern sky; a rocket launched from Australia will not be able to see sources in the northern sky.

Balloons

HIREGS attached to launch vehicle while balloon was being inflated. The instrument has been on three Antarctic campaigns. Balloon flights can carry instruments to altitudes of up to 40 kilometers above sea level, where they are above as much as 99.997 percent of the Earth's atmosphere. Unlike a rocket, which can collect data during a brief few minutes, balloons are able to stay aloft much longer.

However, even at such altitudes, much of the X-ray spectrum is still absorbed by the atmosphere. X-rays with energies less than 35 keV (5,600 aJ) cannot reach balloons. One of the recent balloon-borne experiments was performed by using the High Resolution Gamma-ray and Hard X-ray Spectrometer (HIREGS). It was first launched from McMurdo Station, Antarctica, in December 1991, when steady winds carried the balloon on a circumpolar flight lasting about two weeks. The instrument has been on three Antarctic campaigns.

HIREGS attached to launch vehicle while balloon was being inflated. The instrument has been on three Antarctic campaigns.

Satellites

A detector is placed on a satellite that is then put into orbit well above the Earth's atmosphere. Unlike balloons, instruments on satellites are able to observe the full range of the X-ray spectrum. Unlike sounding rockets, they can collect data for as long as the instruments continue to operate. In one instance, the Vela 5B satellite, the X-ray detector remained functional for over ten years.

Satellites in use today include the XMM-Newton observatory (for low- to mid-energy X-rays, 0.1-15 keV) and the INTEGRAL satellite (high-energy X-rays, 15-60 keV). Both these were launched by the European Space Agency. NASA has launched the Rossi X-ray Timing Explorer (RXTE), and the Swift and Chandra observatories. One of the instruments on Swift is the Swift X-Ray Telescope (XRT). Also, SMART-1 contained an X-ray telescope for mapping lunar X-ray fluorescence. Past observatories included ROSAT, the Einstein Observatory, the ASCA observatory, and BeppoSAX.

X-ray Detectors

CCDs

Most existing X-ray telescopes use CCD (charge-coupled device) detectors, similar to those in visible-light cameras. In visible light, a single photon can produce a single electron of charge in a pixel, and an image is built up by accumulating many such charges from many photons during the exposure time. When an X-ray photon hits a CCD, it produces enough charge (hundreds to thousands of electrons, proportional to its energy) that the individual X-rays have their energies measured on read-out.

Microcalorimeters

Microcalorimeters can detect X-rays only one photon at a time. This works well for astronomical uses, because there just aren't a lot of X-ray photons coming our way, even from the strongest sources like black holes.

Transition Edge Sensors (TES)

TES devices are the next step in microcalorimetery. In essence they are superconducting metals kept as close as possible to their transition temperature, that is, the temperature at which these metals become superconductors and their resistance drops to zero. These transition temperatures are usually just a few degrees above absolute zero (usually less than ten K).

Astronomical Sources of X-rays

Discovery of the first cosmic X-ray source (beyond the Solar System) came as a surprise in 1962. This source is called Scorpius X-1, the first X-ray source found in the constellation of Scorpius, located in the direction of the center of the Milky Way. Based on this discovery, Riccardo Giacconi received the Nobel Prize in Physics in 2002. It was later found that the X-ray emission from this source is 10,000 times greater than its optical emission. In addition, the energy output in X-rays is 100,000 times greater than the total emission of the Sun at all wavelengths.

By now, astronomers have discovered X-rays emissions from several different types of astrophysical objects. These sources include galaxy clusters, black holes in active galactic nuclei (AGN), galactic objects such as supernova remnants, stars, binary stars containing a white dwarf (cataclysmic variable stars), and neutron stars. Some Solar System bodies also emit X-rays, the most notable being the Moon, although most of the X-ray brightness of the Moon arises from reflected solar X-rays. A combination of many unresolved X-ray sources is thought to produce the observed X-ray background, which is occulted by the dark side of the Moon.

Explanation of X-ray Emissions

It is thought that black holes give off radiation because matter falling into them loses gravitational energy, which may result in the emission of radiation before the matter falls into the event horizon. The infalling matter has angular momentum, which means that the material cannot fall in directly, but spins around the black hole. This material often forms an accretion disk. Similar luminous accretion disks can also form around white dwarfs and neutron stars, but in these cases, the infalling matter releases additional energy as it slams against the high-density surface with high speed. In the case of a neutron star, the infalling speed can be a sizable fraction of the speed of light.

In some neutron star or white dwarf systems, the star's magnetic field is strong enough to prevent the formation of an accretion disc. The material in the disc gets very hot because of friction and emits X-rays. The material in the disc slowly loses its angular momentum and falls into the compact star. In the case of neutron stars and white dwarfs, additional X-rays are generated when the material hits their surfaces. X-ray emission from black holes is variable, varying in luminosity in very short timescales. The variation in luminosity can provide information about the size of the black hole.

Clusters of galaxies are formed by the merger of smaller units of matter, such as galaxy groups or individual galaxies. The infalling material (which contains galaxies, gas, and dark matter) gains kinetic energy as it falls into the cluster's gravitational potential well. The infalling gas collides with gas already in the cluster and is shock heated to between 107 and 108 K, depending on the size of the cluster. This very hot material emits X-rays by thermal bremsstrahlung emission, and line emission from "metals." (In astronomy, "metals" often means all elements except hydrogen and helium).

X-rays of Solar System bodies are generally produced by fluorescence. Scattered solar X-rays provide an additional component.

GAMMA-RAY ASTRONOMY

Gamma-ray astronomy is a branch of astronomy that deals with the detection and study of gamma rays in the cosmos. Gamma rays are the most energetic form of electromagnetic radiation and are produced by phenomena such as explosions and high-speed collisions. Thus, gamma-ray astronomy provides the opportunity for scientists to study objects and phenomena in the universe that are associated with extremely high energies.

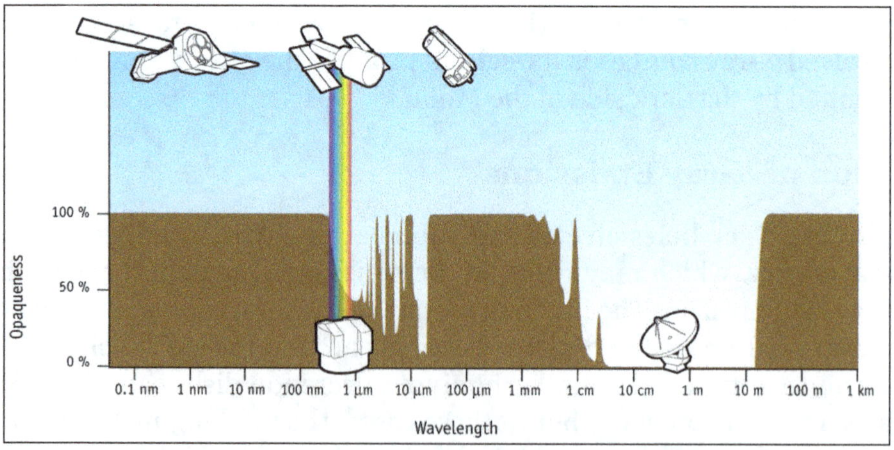

Space telescopes are useful for the study of cosmic gamma rays, most of which are absorbed by the Earth's atmosphere.

Astronomers have found that the universe contains a uniform background of gamma radiation, which has been attributed to the interaction of cosmic rays with interstellar gas. They have also found gamma rays originating from solar flares. In addition, they have detected immensely powerful gamma-ray bursts (GRBs) that appear to come from sources in deep space. These GRBs last anywhere from fractions of a second to minutes, popping off like cosmic flashbulbs from unexpected directions before fading away. Studied for over 25 years with instruments aboard various satellites and space probes, the sources of these high-energy flashes remain a mystery.

Gamma-rays have the smallest wavelengths and the most energy of any other wave in the electromagnetic spectrum. These waves are generated by radioactive atoms and in nuclear explosions. Gamma-rays can kill living cells, a fact which medicine uses to its advantage, using gamma-rays to kill cancerous cells.

Gamma-rays travel to us across vast distances of the universe, only to be absorbed by the Earth's atmosphere. Different wavelengths of light penetrate the Earth's atmosphere to different depths. Instruments aboard high-altitude balloons and satellites like the Compton Observatory provide our only view of the gamma-ray sky.

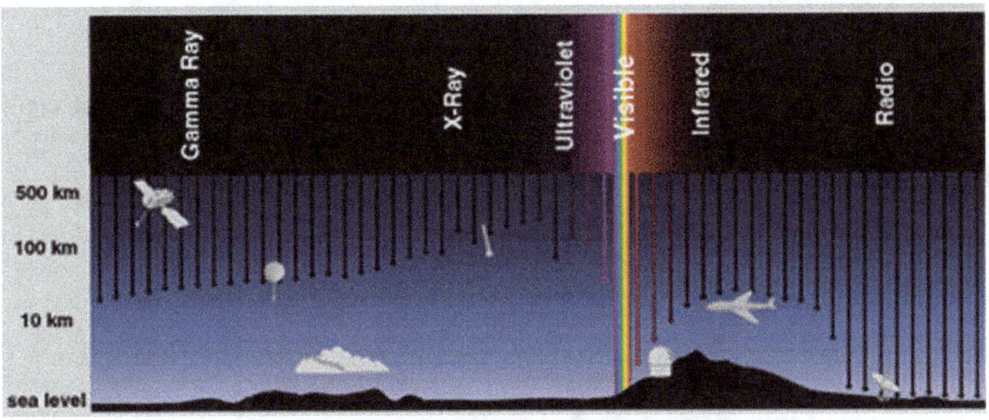

Gamma-rays are the most energetic form of light and are produced by the hottest regions of the universe. They are also produced by such violent events as supernova explosions or the destruction of atoms, and by less dramatic events, such as the decay of radioactive material in space. Things like supernova explosions (the way massive stars die), neutron stars and pulsars, and black holes are all sources of celestial gamma-rays.

Gamma-ray astronomy did not develop until it was possible to get our detectors above all or most of the atmosphere, using balloons or spacecraft. The first gamma-ray telescope, carried into orbit on the Explorer XI satellite in 1961, picked up fewer than 100 cosmic gamma-ray photons.

Unlike optical light and X-rays, gamma rays cannot be captured and reflected in mirrors. The high-energy photons would pass right through such a device. Gamma-ray telescopes use a process called Compton scattering, where a gamma-ray strikes an electron and loses energy, similar to a cue ball striking an eight ball.

This image shows the CGRO satellite being deployed from the Space Shuttle orbiter. This picture was taken from an orbiter window. The two round protrusions are one of CGRO's instruments, called "EGRET".

What do Gamma-rays Show us?

If you could see gamma-rays, the night sky would look strange and unfamiliar.

The gamma-ray moon just looks like a round blob - lunar features are not visible. In high-energy gamma rays, the Moon is actually brighter than the quiet Sun. This image was taken by EGRET.

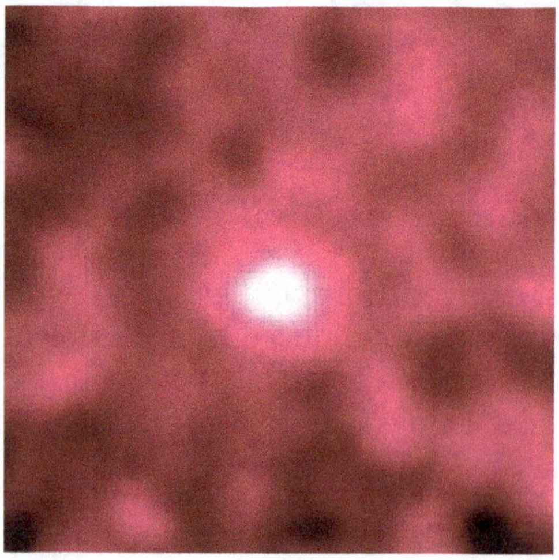

The familiar sights of constantly shining stars and galaxies would be replaced by something ever-changing. Your gamma-ray vision would peer into the hearts of solar flares, supernovae, neutron stars, black holes, and active galaxies. Gamma-ray astronomy presents unique opportunities to explore these exotic objects. By exploring the universe at these high energies, scientists can search for new physics, testing theories and performing experiments which are not possible in earth-bound laboratories.

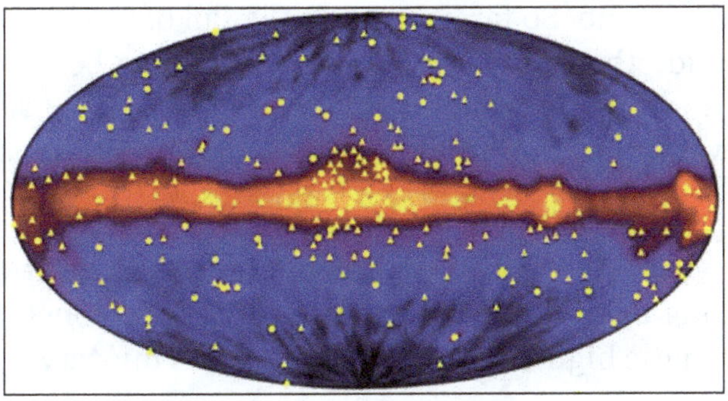

If you could see gamma-rays, these two spinning neutron stars or pulsars would be among the brightest objects in the sky. This computer processed image shows the Crab Nebula pulsar and the Geminga pulsar in the "light" of gamma-rays.

The Crab nebula, shown also in the visible light image, was created by a supernova that brightened the night sky in 1054 A.D. In 1967, astronomers detected the remnant core of that star; a rapidly rotating, magnetic pulsar flashing every 0.33 second in radio waves.

Perhaps the most spectacular discovery in gamma-ray astronomy came in the late 1960s and early 1970s. Detectors on board the Vela satellite series, originally military satellites, began to record bursts of gamma-rays -- not from Earth, but from deep space.

Today, these gamma-ray bursts, which happen at least once a day, are seen to last for fractions of a second to minutes, popping off like cosmic flashbulbs from unexpected directions, flickering, and then fading after briefly dominating the gamma-ray sky.

Gamma-ray bursts can release more energy in 10 seconds than the Sun will emit in its entire 10 billion-year lifetime. So far, it appears that all of the bursts we have observed have come from outside the Milky Way Galaxy. Scientists believe that a gamma-ray burst will occur once every few million years here in the Milky Way, and in fact may occur once every several hundred million years within a few thousand light-years of Earth.

Studied for over 25 years now with instruments on board a variety of satellites and space probes, including Soviet Venera spacecraft and the Pioneer Venus Orbiter, the sources of these enigmatic high-energy flashes remain a mystery.

By solving the mystery of gamma-ray bursts, scientists hope to gain further knowledge of the origins of the Universe, the rate at which the Universe is expanding, and the size of the Universe.

References

- The history of the telescope Henry C. King, Harold Spencer Jones Publisher Courier Dover Publications ISBN 0-486-43265-3, ISBN 978-0-486-43265-6
- Radio-astronomy, astronomy-resources: skyandtelescope.com, Retrieved 14 June, 2020
- Infrared-astronomy, entry: newworldencyclopedia.org, Retrieved 28 April, 2020
- Radio-telescopes, telescopes: nrao.edu, Retrieved 19 May, 2020
- Spiegel, d. S.; burrows, a.; milsom, j. A. (2011). "the deuterium-burning mass limit for brown dwarfs and giant planets". The astrophysical journal. 727 (1): 57. Arxiv:1008.5150
- Cosmochemistry, planetary-science-3: planetary-science.org, Retrieved 28 April, 2020
- "launching the alien debates (part 1 of 7)". Astrobiology magazine. Nasa. 8 december 2006. Retrieved 5 may, 2014
- .X-ray-astronomy, entry: newworldencyclopedia.org, Retrieved 13 July, 2020

Solar Astronomy: Study of the Sun

CHAPTER 3

Solar astronomy is the branch of astronomy that is concerned with the study of sun and its phenomena. Solar physics, solar irradiance, standard solar model, etc. are some of the aspects that come within it. This chapter closely examines solar astronomy to provide an easy understanding of solar astronomy.

Solar astronomy is the scientific study of the celestial star we refer to as the sun and the phenomena that surround it. The main concerns of solar astronomy are the physics, chemistry, meteorology, motion and evolution of the sun as well as its place in the universe.

Historically, astronomy is one of the Earth's oldest sciences. In early civilizations astronomers would perform methodical observations of the night sky, but it was not until the invention of the telescope that astronomy was able to grow and develop into the science that we know today. Because the sun is the physically closest star to the Earth, which is about eight light-minutes away, it is the most frequently studied star.

Our sun is categorized as a typical yellow main-sequence dwarf star of the stellar classification G2 V and is said to be about 4.6 billion years old. The sun makes up about 99% of the total mass of our solar system.

Through the study of solar astronomy scientists have also discovered that the sun goes through an 11-year cycle of periodic changes in activity which they call a sunspot cycle. A sunspot is a region on the sun that has intense magnetic activity which forms an area with lower than average surface temperature. During a sunspot cycle the number of sunspots on the surface of the sun fluctuates.

Solar astronomers have also discovered that the sun has slowly but steadily increased its luminosity by 40% since it first became a main-sequence star. There have also been changes in the sun's luminosity that have directly affected the Earth.

The solar astronomer Edward W. Maunder discovered that during the period between 1645 to 1715 sunspot were exceedingly rare. In that time period astronomers observed only about 50 to 60 sunspots where there are usually around 40,000 to 50,000 sunspots. This phenomenon was named the Maunder Minimum and it was noted that is occurred in the middle of the "Little Ice Age", a time which much of the world was subjected to extremely harsh winters. Whether or not the sunspot activity and the cold winters were merely a coincidence is still an ongoing debate in the scientific community.

Although the sun does not have standard boundaries like a rocky planet such as Earth, it does have a well defined interior structure which is made up of extremely hot gases and surrounded by the photosphere. Because the photosphere is too cool and too thin to radiate a significant amount of light, the surface of the photosphere is the most easily visible part of the sun to the naked eye.

The layer above the photosphere is known as the solar atmosphere which consists of four parts the chromosphere, the transition region, the corona and the heliosphere. The heliosphere while considered the weakest part of the sun's atmosphere extends all the way out past the orbit of Pluto.

SUN

The Sun is the star at the center of the Earth's solar system. The Earth and other matter (including other planets, asteroids, comets, meteoroids, and dust) orbit the Sun, which by itself accounts for more than 99 percent of the solar system's mass. Energy from the Sun—in the form of insolation from sunlight—supports almost all life on Earth via photosynthesis, and drives the Earth's climate and weather.

About 74 percent of the Sun's mass is hydrogen, 25 percent is helium, and the rest is made up of trace quantities of heavier elements. The Sun is thought to be about 4.6 billion years old and about halfway through its main-sequence evolution. Within the Sun's core, nuclear fusion reactions take place, with hydrogen nuclei being fused into helium nuclei. Through these reactions, more than 4 million tons of matter are converted into energy each second, producing neutrinos and solar radiation. Current theory predicts that in about five billion years, the Sun will evolve into a red giant and then a white dwarf, creating a planetary nebula in the process.

The Sun is a magnetically active star. It supports a strong, changing magnetic field that varies year-to-year and reverses direction about every 11 years. The Sun's magnetic field gives rise to many effects that are collectively called solar activity. They include sunspots on the Sun's surface, solar flares, and variations in the solar wind that carry material through the solar system. The effects of solar activity on Earth include auroras at moderate to high latitudes, and the disruption of radio communications and electric power. Solar activity is thought to have played a large role in the formation and evolution of the solar system, and strongly affects the structure of the Earth's outer atmosphere.

Although it is the nearest star to Earth and has been intensively studied by scientists, many questions about the Sun remain unanswered. For instance, we do not know why its outer atmosphere has a temperature of over a million K while its visible surface (the photosphere) has a temperature of just 6,000 K. Current topics of scientific inquiry include the Sun's regular cycle of sunspot activity, the physics and origin of solar flares and prominences, the magnetic interaction between the chromosphere and the corona, and the origin of the solar wind.

Its astrological and astronomical symbol is a circle with a point at its center: ⊙ Some ancient peoples of the world considered it a planet.

The Sun is placed in a spectral class called G2V. "G2" means that it has a surface temperature of approximately 5,500 K, giving it a white color. As a consequence of light scattering by the Earth's atmosphere, it appears yellow to us. Its spectrum contains lines of ionized and neutral metals, as well as very weak hydrogen lines. The "V" suffix indicates that the Sun, like most stars, is a main sequence star. This means that it generates its energy by nuclear fusion of hydrogen nuclei into helium and is in a state of hydrostatic balance—neither contracting nor expanding over time. There are more than 100 million G2 class stars in our galaxy. Due to logarithmic size distribution, the Sun is actually brighter than 85 percent of the stars in the Galaxy, most of which are red dwarfs.

The Sun as it appears through a camera lens from the Earth's surface.

The Sun will spend a total of approximately 10 billion years as a main sequence star. Its current age, determined using computer models of stellar evolution and nucleocosmochronology, is thought to be about 4.57 billion years. The Sun orbits the center of the Milky Way galaxy at a distance of about 25,000 to 28,000 light-years from the galactic center, completing one revolution in about 225–250 million years. The orbital speed is 220 km/s, equivalent to one light-year every 1,400 years, and one AU every 8 days.

It is suggested that the Sun is a third generation star, whose formation may have been triggered by shockwaves from a nearby supernova based on a high abundance of heavy elements such as gold and uranium in the solar system. These elements could most plausibly have been produced by endergonic nuclear reactions during a supernova, or by transmutation via neutron absorption inside a massive second-generation star.

The Sun does not have enough mass to explode as a supernova. Instead, in 4–5 billion years, it will enter a red giant phase, its outer layers expanding as the hydrogen fuel in the core is consumed and the core contracts and heats up. Helium fusion will begin when the core temperature reaches about 3×108 K. While it is likely that the expansion of the outer layers of the Sun will reach the current position of Earth's orbit, recent research suggests that mass lost from the Sun earlier in its red giant phase will cause the Earth's orbit to move further out, preventing it from being engulfed. However, Earth's water and most of the atmosphere will be boiled away.

Following the red giant phase, intense thermal pulsations will cause the Sun to throw off its outer layers, forming a planetary nebula. The Sun will then evolve into a white dwarf, slowly cooling over eons. This stellar evolution scenario is typical of low- to medium-mass stars.

Sunlight is the main source of energy near the surface of Earth. The solar constant is the amount of power that the Sun deposits per unit area that is directly exposed to sunlight. The solar constant is equal to approximately 1,370 watts per square meter of area at a distance of one AU from the Sun (that is, on or near Earth). Sunlight on the surface of Earth is attenuated by the Earth's atmosphere so that less power arrives at the surface—closer to 1,000 watts per directly exposed square meter in clear conditions when the Sun is near the zenith. This energy can be harnessed via a variety of natural and synthetic processes—photosynthesis by plants captures the energy of sunlight and converts it to chemical form (oxygen and reduced carbon compounds), while direct heating or electrical conversion by solar cells are used by solar power equipment to generate electricity or to do other useful work. The energy stored in petroleum and other fossil fuels was originally converted from sunlight by photosynthesis in the distant past.

Sunlight has several interesting biological properties. Ultraviolet light from the Sun has antiseptic properties and can be used to sterilize tools. It also causes sunburn, and has other medical effects such as the production of Vitamin D. Ultraviolet light is strongly attenuated by Earth's atmosphere, so that the amount of UV varies greatly with latitude due the longer passage of sunlight through the atmosphere at high latitudes. This variation is responsible for many biological adaptations, including variations in human skin color in different regions of the globe.

Observed from Earth, the path of the Sun across the sky varies throughout the year. The shape described by the Sun's position, considered at the same time each day for a complete year, is called the analemma and resembles a figure aligned along a North/South axis. While the most obvious variation in the Sun's apparent position through the year is a North/South swing over 47 degrees of angle (due to the 23.5-degree tilt of the Earth with respect to the Sun), there is an East/West component as well. The North/South swing in apparent angle is the main source of seasons on Earth.

Theoretical Problems

Solar Neutrino Problem

For many years the number of solar electron neutrinos detected on Earth was only a third of the number expected, according to theories describing the nuclear reactions in the Sun. This anomalous result was termed the solar neutrino problem. Theories proposed to resolve the problem either tried to reduce the temperature of the Sun's interior to explain the lower neutrino flux, or posited that electron neutrinos could oscillate, that is, change into undetectable tau and muon neutrinos as they traveled between the Sun and the Earth. Several neutrino observatories were built in the 1980s to measure the solar neutrino flux as accurately as possible, including the Sudbury Neutrino Observatory and Kamiokande. Results from these observatories eventually led to the discovery that neutrinos have a very small rest mass and can indeed oscillate. Moreover, the Sudbury Neutrino Observatory was able to detect all three types of neutrinos directly, and found that the Sun's total neutrino emission rate agreed with the Standard Solar Model, although only one-third of the neutrinos seen at Earth were of the electron type.

Coronal Heating Problem

The optical surface of the Sun (the photosphere) is known to have a temperature of approximately 6,000 K. Above it lies the solar corona at a temperature of 1,000,000 K. The high temperature of the corona shows that it is heated by something other than the photosphere.

It is thought that the energy necessary to heat the corona is provided by turbulent motion in the convection zone below the photosphere, and two main mechanisms have been proposed to explain coronal heating. The first is wave heating, in which sound, gravitational and magnetohydrodynamic waves are produced by turbulence in the convection zone. These waves travel upward and dissipate in the corona, depositing their energy in the ambient gas in the form of heat. The other is magnetic heating, in which magnetic energy is continuously built up by photospheric motion and released through magnetic reconnection in the form of large solar flares and myriad similar but smaller events.

Currently, it is unclear whether waves are an efficient heating mechanism. All waves except Alfven waves have been found to dissipate or refract before reaching the corona. In addition, Alfven waves do not easily dissipate in the corona. Current research focus has therefore shifted towards flare heating mechanisms. One possible candidate to explain coronal heating is continuous flaring at small scales, but this remains an open topic of investigation.

Faint Young Sun Problem

Theoretical models of the sun's development suggest that 3.8 to 2.5 billion years ago, during the Archean period, the Sun was only about 75% as bright as it is today. Such a weak star would not have been able to sustain liquid water on the Earth's surface, and thus life should not have been able to develop. However, the geological record demonstrates that the Earth has remained at a fairly constant temperature throughout its history, and in fact that the young Earth was somewhat warmer than it is today. The general consensus among scientists is that the young Earth's atmosphere contained much larger quantities of greenhouse gases (such as carbon dioxide and/or ammonia) than are present today, which trapped enough heat to compensate for the lesser amount of solar energy reaching the planet.

Magnetic Field

The heliospheric current sheet extends to the outer reaches of the Solar System, and results from the influence of the Sun's rotating magnetic field on the plasma in the interplanetary medium.

All matter in the Sun is in the form of gas and plasma due to its high temperatures. This makes it possible for the Sun to rotate faster at its equator (about 25 days) than it does at higher latitudes (about 35 days near its poles). The differential rotation of the Sun's latitudes causes its magnetic field lines to become twisted together over time, causing magnetic field loops to erupt from the Sun's surface and trigger the formation of the Sun's dramatic sunspots and solar prominences. This twisting action gives rise to the solar dynamo and an 11-year solar cycle of magnetic activity as the Sun's magnetic field reverses itself about every 11 years.

The influence of the Sun's rotating magnetic field on the plasma in the interplanetary medium creates the heliospheric current sheet, which separates regions with magnetic fields pointing in different directions. The plasma in the interplanetary medium is also responsible for the strength of the Sun's magnetic field at the orbit of the Earth. If space were a vacuum, then the Sun's 10^{-4} tesla magnetic dipole field would reduce with the cube of the distance to about 10^{-11} tesla. But satellite observations show that it is about 100 times greater at around 10^{-9} tesla. Magnetohydrodynamic (MHD) theory predicts that the motion of a conducting fluid (e.g., the interplanetary medium) in a magnetic field, induces electric currents which in turn generates magnetic fields, and in this respect it behaves like an MHD dynamo.

Sun Observation and Eye Damage

Sunlight is very bright, and looking directly at the Sun with the naked eye for brief periods can be painful, but is generally not hazardous. Looking directly at the Sun causes phosphene visual artifacts and temporary partial blindness. It also delivers about 4 milliwatts of sunlight to the retina, slightly heating it and potentially (though not normally) damaging it. UV exposure gradually yellows the lens of the eye over a period of years and can cause cataracts, but those depend on general exposure to solar UV, not on whether one looks directly at the Sun.

Viewing the Sun through light-concentrating optics such as binoculars is very hazardous without an attenuating (ND) filter to dim the sunlight. Using a proper filter is important as some improvised filters pass UV rays that can damage the eye at high brightness levels. Unfiltered binoculars can deliver over 500 times more sunlight to the retina than does the naked eye, killing retinal cells almost instantly. Even brief glances at the midday Sun through unfiltered binoculars can cause permanent blindness. One way to view the Sun safely is by projecting an image onto a screen using binoculars or a small telescope.

Partial solar eclipses are hazardous to view because the eye's pupil is not adapted to the unusually high visual contrast: the pupil dilates according to the total amount of light in the field of view, not by the brightest object in the field. During partial eclipses most sunlight is blocked by the Moon passing in front of the Sun, but the uncovered parts of the photosphere have the same surface brightness as during a normal day. In the overall gloom, the pupil expands from ~2 mm to ~6 mm, and each retinal cell exposed to the solar image receives about ten times more light than it would looking at the non-eclipsed sun. This can damage or kill those cells, resulting in small permanent blind spots for the viewer. The hazard is insidious for inexperienced observers and for children, because there is no perception of pain: it is not immediately obvious that one's vision is being destroyed.

During sunrise and sunset, sunlight is attenuated through rayleigh and mie scattering of light by a particularly long passage through Earth's atmosphere, and the direct Sun is sometimes faint enough to be viewed directly without discomfort or safely with binoculars. Hazy conditions, atmospheric dust, and high humidity contribute to this atmospheric attenuation.

Solar Core

The core of the Sun is considered to extend from the center to about 0.2 to 0.25 of solar radius. It is the hottest part of the Sun and of the Solar System. It has a density of 150 g/cm^3 (150 times the density of liquid water) at the center, and a temperature of 15 million kelvins (15 million degrees Celsius, 27 million degrees Fahrenheit). The core is made of hot, dense plasma (ions and electrons), at a pressure estimated at 265 billion bar (3.84 trillion psi or 26.5 petapascals (PPa)) at the center. Due to fusion, the composition of the solar plasma drops from 68–70% hydrogen by mass at the outer core, to 33% hydrogen at the core/Sun center.

The core inside 0.20 of the solar radius contains 34% of the Sun's mass, but only 3.4% of the Sun's volume. Inside the 0.24 solar radius is the core which generates 99% of the fusion power of the Sun. There are two distinct reactions in which four hydrogen nuclei may eventually result in one helium nucleus: the proton-proton chain reaction – which is responsible for most of the Sun's released energy – and the CNO cycle.

Composition

The Sun at the photosphere is about 73–74% by mass hydrogen, which is the same composition as the atmosphere of Jupiter, and the primordial composition of hydrogen and helium at the earliest star formation after the Big Bang. However, as depth into the Sun increases, fusion decreases the fraction of hydrogen. Traveling inward, hydrogen mass fraction starts to decrease rapidly after the core radius has been reached (it is still about 70% at a radius 25% of the Sun's radius) and inside this, the hydrogen fraction drops rapidly as the core is traversed, until it reaches a low of about 33% hydrogen, at the Sun's center (radius zero). All but 2% of the remaining plasma mass (i.e., 65%) is helium, at the center of the Sun.

Energy Conversion

Approximately 3.7×10^{38} protons (hydrogen nuclei), or roughly 600 million metric tons of hydrogen, are converted into helium nuclei every second releasing energy at a rate of 3.86×10^{26} joules per second.

The core produces almost all of the Sun's heat via fusion: the rest of the star is heated by the outward transfer of heat from the core. The energy produced by fusion in the core, except a small part carried out by neutrinos, must travel through many successive layers to the solar photosphere before it escapes into space as sunlight, kinetic or thermal energy of particles. The energy conversion per unit time (power) of fusion in the core varies with distance from the solar center. At the center of the Sun, fusion power is estimated by models to be about 276.5 watts/m^3. Despite its intense temperature, the peak power generating density of the core overall is similar to an active compost heap, and is lower than the power density produced by the metabolism of an adult human. The Sun is much hotter than a compost heap due to the Sun's enormous volume.

The low power outputs occurring inside the fusion core of the Sun may also be surprising, considering the large power which might be predicted by a simple application of the Stefan–Boltzmann law for temperatures of 10 to 15 million kelvins. However, layers of the Sun are radiating to outer layers only slightly lower in temperature, and it is this difference in radiation powers between layers which determines net power generation and transfer in the solar core.

At 19% of the solar radius, near the edge of the core, temperatures are about 10 million kelvins and fusion power density is 6.9 W/m³, which is about 2.5% of the maximum value at the solar center. The density here is about 40 g/cm³, or about 27% of that at the center. Some 91% of the solar energy is produced within this radius. Within 24% of the radius (the outer "core" by some definitions), 99% of the Sun's power is produced. Beyond 30% of the solar radius, where temperature is 7 million K and density has fallen to 10 g/cm³ the rate of fusion is almost nil. There are two distinct reactions in which 4 H nuclei may eventually result in one He nucleus: "proton-proton chain reaction" and the "CNO cycle".

Proton-proton Chain Reaction

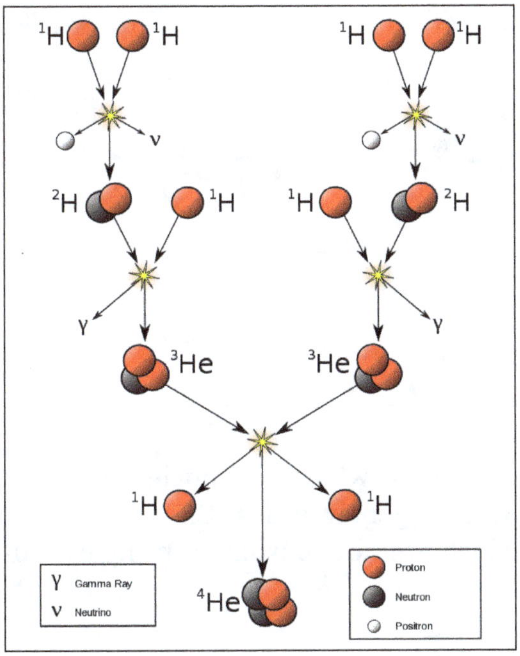

The first reaction in which 4 H nuclei may eventually result in one He nucleus is known as the proton-proton chain, is:

$$\begin{cases} & {}^{1}\mathrm{H} + {}^{1}\mathrm{H} \rightarrow {}^{2}\mathrm{D} + e^{+} + \nu_e \\ \text{then} & {}^{2}\mathrm{D} + {}^{1}\mathrm{H} \rightarrow {}^{3}\mathrm{He} + \gamma \\ \text{then} & {}^{3}\mathrm{He} + {}^{3}\mathrm{He} \rightarrow {}^{4}\mathrm{He} + {}^{1}\mathrm{H} + {}^{1}\mathrm{H} \end{cases}$$

This reaction sequence is thought to be the most important one in the solar core. The characteristic time for the first reaction is about one billion years even at the high densities and temperatures of the core, due to the necessity for the weak force to cause beta decay before the nucleons can adhere (which rarely happens in the time they tunnel toward each other, to be close enough to do so). The time that deuterium and helium-3 in the next reactions last, by contrast, are only about 4 seconds and 400 years. These later reactions proceed via the nuclear force and are thus much faster. The total energy released by these reactions in turning 4 hydrogen atoms into 1 helium atom is 26.7 MeV.

CNO Cycle

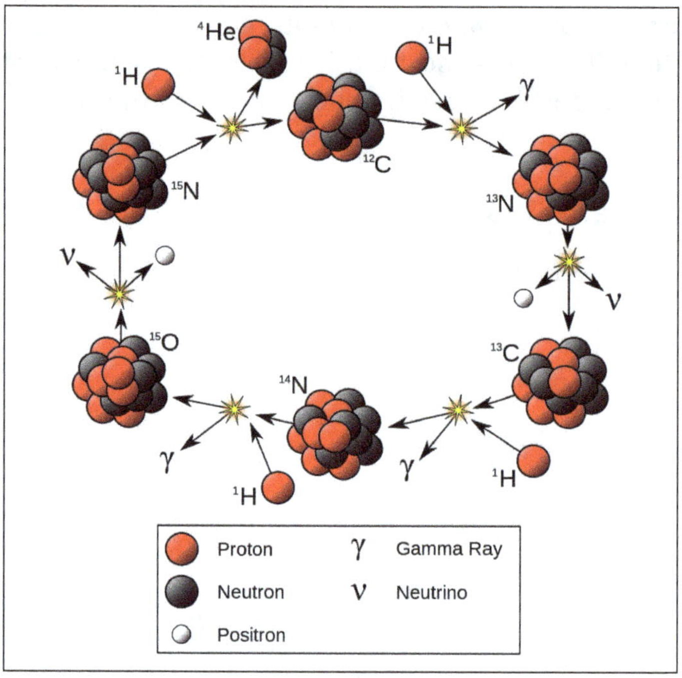

The second reaction sequence, in which 4 H nuclei may eventually result in one He nucleus is called the CNO cycle generates less than 10% of the total solar energy. This involves carbon atoms which are not consumed in the overall process. The details of this CNO cycle are as follows:

$$\begin{cases} {}^{12}C + {}^{1}H \rightarrow {}^{13}N + \gamma \\ \text{then} \quad {}^{13}N \rightarrow {}^{13}C + e^{+} + \nu_e \\ \text{then} \quad {}^{13}C + {}^{1}H \rightarrow {}^{14}N + \gamma \\ \text{then} \quad {}^{14}N + {}^{1}H \rightarrow {}^{15}O + \gamma \\ \text{then} \quad {}^{15}O \rightarrow {}^{15}N + e^{+} + \nu_e \\ \text{then} \quad {}^{15}N + {}^{1}H \rightarrow {}^{12}C + {}^{4}He + \gamma \end{cases}$$

This process can be further understood by the picture above, starting from the top in clockwise direction.

Equilibrium

The rate of nuclear fusion depends strongly on density. Therefore, the fusion rate in the core is in a self-correcting equilibrium: a slightly higher rate of fusion would cause the core to heat up more and expand slightly against the weight of the outer layers. This would reduce the fusion rate and correct the perturbation; and a slightly lower rate would cause the core to cool and shrink slightly, increasing the fusion rate and again reverting it to its present level.

However the Sun gradually becomes hotter during its time on the main sequence, because the helium atoms in the core are denser than the hydrogen atoms they were fused from. This increases the gravitational pressure on the core which is resisted by a gradual increase in the rate at which fusion occurs. This process speeds up over time as the core gradually becomes denser. It is estimated that the Sun has become 30% brighter in the last four and a half billion years and will continue to increase in brightness by 1% every 100 million years.

Energy Transfer

The high-energy photons (gamma rays) released in fusion reactions take indirect paths to the Sun's surface. According to current models, random scattering from free electrons in the solar radiative zone (the zone within 75% of the solar radius, where heat transfer is by radiation) sets the photon diffusion time scale (or "photon travel time") from the core to the outer edge of the radiative zone at about 170,000 years. From there they cross into the convective zone (the remaining 25% of distance from the Sun's center), where the dominant transfer process changes to convection, and the speed at which heat moves outward becomes considerably faster.

In the process of heat transfer from core to photosphere, each gamma ray in the Sun's core is converted during scattering into several million visible light photons before escaping into space. Neutrinos are also released by the fusion reactions in the core, but unlike photons they very rarely interact with matter, so almost all are able to escape the Sun immediately. For many years measurements of the number of neutrinos produced in the Sun were much lower than theories predicted, a problem which was recently resolved through a better understanding of neutrino oscillation.

Radiation Zone

A radiation zone, or radiative region is a layer of a star's interior where energy is primarily transported toward the exterior by means of radiative diffusion and thermal conduction, rather than by convection. Energy travels through the radiation zone in the form of electromagnetic radiation as photons.

Matter in a radiation zone is so dense that photons can travel only a short distance before they are absorbed or scattered by another particle, gradually shifting to longer wavelength as they do so. For this reason, it takes an average of 171,000 years for gamma rays from the core of the Sun to leave the radiation zone. Over this range, the temperature of the plasma drops from 15 million K near the core down to 1.5 million K at the base of the convection zone.

Temperature Gradient

In a radiative zone, the temperature gradient—the change in temperature (T) as a function of radius (r)—is given by:

$$\frac{dT(r)}{dr} = -\frac{3\kappa(r)\rho(r)L(r)}{(4\pi r^2)(16\sigma_B)T^3(r)}$$

where $\kappa(r)$ is the opacity, $\rho(r)$ is the matter density, $L(r)$ is the luminosity, and σ_B is the Stefan–Boltzmann constant. Hence the opacity (κ) and radiation flux (L) within a given layer of a star are important factors in determining how effective radiative diffusion is at transporting energy. A high opacity or high luminosity can cause a high temperature gradient, which results from a slow flow of energy. Those layers where convection is more effective than radiative diffusion at transporting energy, thereby creating a lower temperature gradient, will become convection zones.

This relation can be derived by integrating Fick's first law over the surface of some radius r, giving the total outgoing energy flux which is equal to the luminosity by conservation of energy:

$$L = -4\pi r^2 D\frac{\partial u}{\partial r}$$

Where D is the photons diffusion coefficient, and u is the energy density.

The energy density is related to the temperature by Stefan-Boltzmann law by:

$$U = \frac{4}{c}\sigma_B T^4$$

Finally, as in the elementary theory of diffusion coefficient in gases, the diffusion coefficient D approximately satisfies:

$$D = \frac{1}{3}c\lambda$$

where λ is the photon mean free path, and is the reciprocal of the opacity κ.

Eddington Stellar Model

Eddington assumed the pressure P in a star is a combination of an ideal gas pressure and radiation pressure, and that there is a constant ratio, β, of the gas pressure to the total pressure. Therefore, by the ideal gas law:

$$\beta P = k_B \frac{\rho}{\mu} T$$

where k_B is Boltzmann constant and μ the mass of a single atom (actually, an ion since matter is ionized; usually a hydrogen ion, i.e. a proton). While the radiation pressure satisfies:

$$1-\beta = \frac{P_{\text{radiation}}}{P} = \frac{u}{3P} = \frac{4\sigma_B}{3c}\frac{T^4}{P}$$

so that T^4 is proportional to P throughout the star.

This gives the polytropic equation (with n=3):

$$P = \left(\frac{3ck_B^4}{4\sigma_B \mu^4}\frac{1-\beta}{\beta^4}\right)^{1/3} \rho^{4/3}$$

Using the hydrostatic equilibrium equation, the second equation becomes equivalent to:

$$-\frac{GM\rho}{r^2} = \frac{\mathrm{d}P}{\mathrm{d}r} = \frac{16\sigma_B}{3c(1-\beta)} T^3 \frac{\mathrm{d}T}{\mathrm{d}r}$$

For energy transmission by radiation only, we may use the equation for the temperature gradient (presented in the previous subsection) for the right-hand side and get:

$$GM = \frac{\kappa L}{4\pi c(1-\beta)}$$

Thus the Eddington model is a good approximation in the radiation zone as long as $\kappa L/M$ is approximately constant, which is often the case.

Stability against Convection

The radiation zone is stable against formation of convection cells if the density gradient is high enough, so that an element moving upwards has its density lowered (due to adiabatic expansion) less than the drop in density of its surrounding, so that it will experience a net buoyancy force downwards.

The criterion for this is:

$$\frac{\mathrm{d}\log\rho}{\mathrm{d}\log P} > \frac{1}{\gamma_{ad}}$$

where P is the pressure, ρ the density and is the heat capacity ratio.

For a homogenic ideal gas, this is equivalent to:

$$\frac{d\log T}{d\log P} < 1 - \frac{1}{\gamma_{ad}}$$

We can calculate the left-hand side by dividing the equation for the temperature gradient by the equation relating the pressure gradient to the gravity acceleration g:

$$\frac{dP(r)}{dr} = g\rho = \frac{GM(r)\rho(r)}{r^2}$$

$M(r)$ being the mass within the sphere of radius r, and is approximately the whole star mass for large enough r.

This gives the following form of the Schwarzschild criterion for stability against convection:

$$\frac{3}{64\pi\sigma_B G}\frac{\kappa L}{M}\frac{P}{T^4} < 1 - \frac{1}{\gamma_{ad}}$$

Note that for non-homogenic gas this criterion should be replaced by the Ledoux criterion, because the density gradient now also depends on concentration gradients.

For a polytrope solution with n=3 (as in the Eddington stellar model for radiation zone), P is proportional to T^4 and the left-hand side is constant and equals 1/4, smaller than the ideal monoatomic gas approximation for the right-hand side giving $1 - 1/\gamma_{ad} = 2/5$. This explains the stability of the radiation zone against convection.

However, at a large enough radius, the opacity κ increases due to the decrease in temperature (by Kramers' opacity law), and possibly also due to a smaller degree of ionization in the lower shells of heavy elements ions. This leads to a violation of the stability criterion and to the creation of the convection zone; in the sun, opacity increases by more than a tenfold across the radiation zone, before the transition to the convection zone happens.

Additional situations in which this stability criterion is not met are:

- Large values of $L(r) / M(r)$, which may happen towards the star core's center, where $M(r)$ is small, if nuclear energy production is strongly peaked at the center, as in relatively massive stars. Thus such stars have a convective core.
- A smaller value of γ_{ad}. For semi-ionized gas, where approximately half of the atoms are ionized, the effective value of γ_{ad} drops to 6/5, giving $1 - 1/\gamma_{ad} = 1/6$. Therefore, all stars have shallow convection zones near their surfaces, at low enough temperatures where ionization is only partial.

Main Sequence Stars

For main sequence stars—those stars that are generating energy through the thermonuclear fusion of hydrogen at the core, the presence and location of radiative regions depends on the star's mass. Main sequence stars below about 0.3 solar masses are entirely convective, meaning they do not have a radiative zone. From 0.3 to 1.2 solar masses, the region around the stellar core is a radiation zone, separated from the overlying convection zone by the tachocline. The radius of the radiative zone increases monotonically with mass, with stars around 1.2 solar masses being almost entirely radiative. Above 1.2 solar masses, the core region becomes a convection zone and the overlying region is a radiation zone, with the amount of mass within the convective zone increasing with the mass of the star.

The Sun

In the Sun, the region between the solar core at 0.2 of the Sun's radius and the outer convection zone at 0.71 of the Sun's radius is referred to as the radiation zone, although the core is also a radiative region. The convection zone and the radiation zone are divided by the tachocline, another part of the Sun.

Convection Zone

A convection zone, convective zone or convective region of a star is a layer which is unstable to convection. Energy is primarily or partially transported by convection in such a region. In a radiation zone, energy is transported by radiation and conduction.

Stellar convection consists of mass movement of plasma within the star which usually forms a circular convection current with the heated plasma ascending and the cooled plasma descending.

The Schwarzschild criterion expresses the conditions under which a region of a star is unstable to convection. A parcel of gas that rises slightly will find itself in an environment of lower pressure than the one it came from. As a result, the parcel will expand and cool. If the rising parcel cools to a lower temperature than its new surroundings, so that it has a higher density than the surrounding gas, then its lack of buoyancy will cause it to sink back to where it came from. However, if the temperature gradient is steep enough (i. e. the temperature changes rapidly with distance from the center of the star), or if the gas has a very high heat capacity (i. e. its temperature changes relatively slowly as it expands) then the rising parcel of gas will remain warmer and less dense than its new surroundings even after expanding and cooling. Its buoyancy will then cause it to continue to rise. The region of the star in which this happens is the convection zone.

Main Sequence Stars

In main sequence stars more than 1.3 times the mass of the Sun, the high core temperature causes nuclear fusion of hydrogen into helium to occur predominantly via the carbon-nitrogen-oxygen (CNO) cycle instead of the less temperature-sensitive proton-proton chain. The high temperature gradient in the core region forms a convection zone that slowly mixes the hydrogen fuel with the helium product. The core convection zone of these stars is overlaid by a radiation zone that is in thermal equilibrium and undergoes little or no mixing. In the most massive stars, the convection zone may reach all the way from the core to the surface.

In main sequence stars of less than about 1.3 solar masses, the outer envelope of the star contains a region where partial ionization of hydrogen and helium raises the heat capacity. The relatively low temperature in this region simultaneously causes the opacity due to heavier elements to be high enough to produce a steep temperature gradient. This combination of circumstances produces an outer convection zone, the top of which is visible in the Sun as solar granulation. Low mass main sequences of stars, such as red dwarfs below 0.35 solar masses, as well as pre-main sequence stars on the Hayashi track, are convective throughout and do not contain a radiation zone.

In main sequence stars similar to the Sun, which have a radiative core and convective envelope, the transition region between the convection zone and the radiation zone is called the tachocline.

Red Giants

In red giant stars, and particularly during the asymptotic giant branch phase, the surface convection zone varies in depth during the phases of shell burning. This causes dredge-up events, short-lived very deep convection zones that transport fusion products to the surface of the star.

Photosphere

The photosphere is a star's outer shell from which light is radiated. It extends into a star's surface until the plasma becomes opaque, equivalent to an optical depth of approximately 2/3, or equivalently, a depth from which 50% of light will escape without being scattered. In other words, a photosphere is the deepest region of a luminous object, usually a star, that is transparent to photons of certain wavelengths.

Temperature

The surface of a star is defined to have a temperature given by the effective temperature in the Stefan–Boltzmann law. Stars, except neutron stars, have no solid or liquid surface. Therefore, the photosphere is typically used to describe the Sun's or another star's visual surface.

Composition of the Sun

The Sun is composed primarily of the chemical elements hydrogen and helium; they account for 74.9% and 23.8% of the mass of the Sun in the photosphere, respectively. All heavier elements, called metals in astronomy, account for less than 2% of the mass, with oxygen (roughly 1% of the Sun's mass), carbon (0.3%), neon (0.2%), and iron (0.2%) being the most abundant.

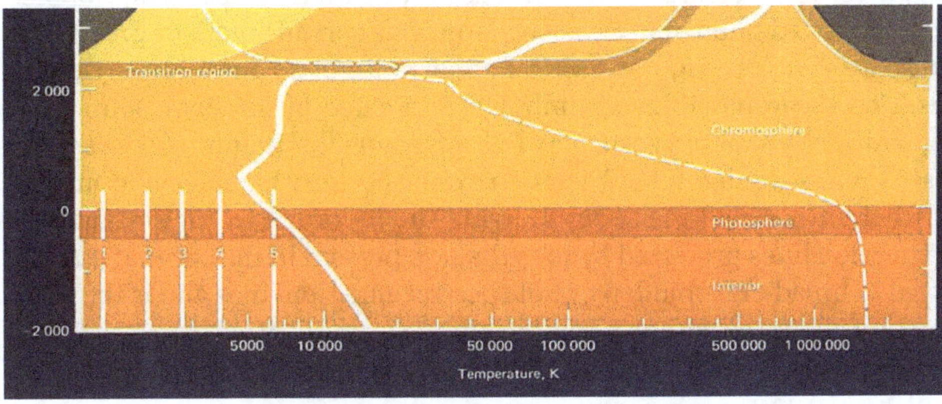

Solar atmosphere: temperature and density.

The Sun's photosphere has a temperature between 4,500 and 6,000 K (4,230 and 5,730 °C) (with an effective temperature of 5,777 K (5,504 °C)) and a density of about 1×10^{-6} kg/m^3; increasing with depth into the sun. Other stars may have hotter or cooler photospheres. The Sun's photosphere is around 100 kilometers thick, and is composed of convection cells called granules—cells of plasma each approximately 1000 kilometers in diameter with hot rising plasma in the center and cooler plasma falling in the narrow spaces between them, flowing at velocities of 7 kilometer per second. Each granule has a lifespan of only about twenty minutes, resulting in a continually shifting "boiling" pattern. Grouping the typical granules are super granules up to 30,000 kilometers in diameter with lifespans of up to 24 hours and flow speeds of about 500 meter per second, carrying magnetic field bundles to the edges of the cells. Other magnetically-related phenomena include sunspots and solar faculae dispersed between the granules. These details are too fine to be seen when observing other stars from earth.

Other Layers of the Sun

The Sun's visible atmosphere has other layers above the photosphere: the 2,000 kilometer-deep chromosphere (typically observed by filtered light, for example H-alpha) lies just between the photosphere and the much hotter but more tenuous corona. Other "surface features" on the photosphere are solar flares and sunspots.

Corona

Corona is an aura of plasma that surrounds the Sun and other stars. The Sun's corona extends millions of kilometres into outer space and is most easily seen during a total solar eclipse, but it is also observable with a coronagraph.

Spectroscopy measurements indicate strong ionization in the corona and a plasma temperature in excess of 1000000 kelvin, much hotter than the surface of the Sun.

Light from the corona comes from three primary sources, from the same volume of space. The K-corona is created by sunlight scattering off free electrons; Doppler broadening of the reflected photospheric absorption lines spreads them so greatly as to completely obscure them, giving the spectral appearance of a continuum with no absorption lines. The F-corona (F for Fraunhofer) is created by sunlight bouncing off dust particles, and is observable because its light contains the Fraunhofer absorption lines that are seen in raw sunlight; the F-corona extends to very high elongation angles from the Sun, where it is called the zodiacal light. The E-corona (E for emission) is due to spectral emission lines produced by ions that are present in the coronal plasma; it may be observed in broad or forbidden or hot spectral emission lines and is the main source of information about the corona's composition.

Physical Features

The sun's corona is much hotter (by a factor from 150 to 450) than the visible surface of the Sun: the photosphere's average temperature is 5800 kelvin compared to the corona's one to three million kelvin. The corona is 10^{-12} times as dense as the photosphere, and so produces about one-millionth as much visible light. The corona is separated from the photosphere by the relatively shallow chromosphere. The exact mechanism by which the corona is heated is still the subject of some debate, but likely possibilities include induction by the Sun's magnetic field and magnetohydrodynamic waves from below. The outer edges of the Sun's corona are constantly being transported away due to open magnetic flux and hence generating the solar wind.

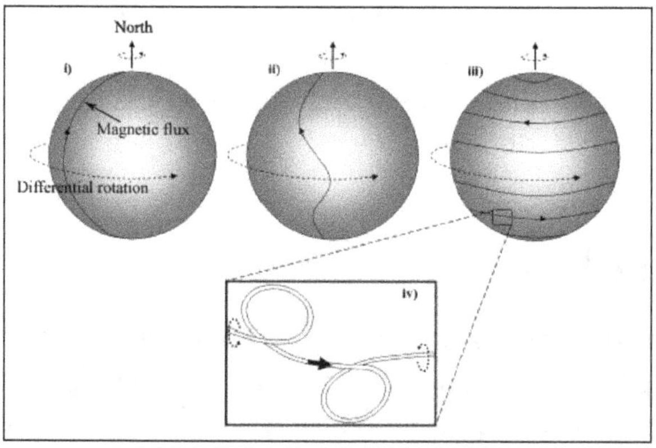

A drawing demonstrating the configuration of solar magnetic flux during the solar cycle.

The corona is not always evenly distributed across the surface of the sun. During periods of quiet, the corona is more or less confined to the equatorial regions, with coronal holes covering the polar regions. However, during the Sun's active periods, the corona is evenly distributed over the equatorial and polar regions, though it is most prominent in areas with sunspot activity. The solar cycle spans approximately 11 years, from solar minimum to the following minimum. Since the solar magnetic field is continually wound up due to the faster rotation of mass at the sun's equator (differential rotation), sunspot activity will be more pronounced at solar maximum where the magnetic field is more twisted. Associated with sunspots are coronal loops, loops of magnetic flux, upwelling from the solar interior. The magnetic flux pushes the hotter photosphere aside, exposing the cooler plasma below, thus creating the relatively dark sun spots.

Since the corona has been photographed at high resolution in the X-ray range of the spectrum by the satellite Skylab in 1973, and then later by Yohkoh and the other following space instruments, it has been seen that the structure of the corona is quite varied and complex: different zones have been immediately classified on the coronal disc. The astronomers usually distinguish several regions, as described below.

Active Regions

Active regions are ensembles of loop structures connecting points of opposite magnetic polarity in the photosphere, the so-called coronal loops. They generally distribute in two zones of activity, which are parallel to the solar equator. The average temperature is between two and four million kelvin, while the density goes from 10^9 to 10^{10} particle per cm^3.

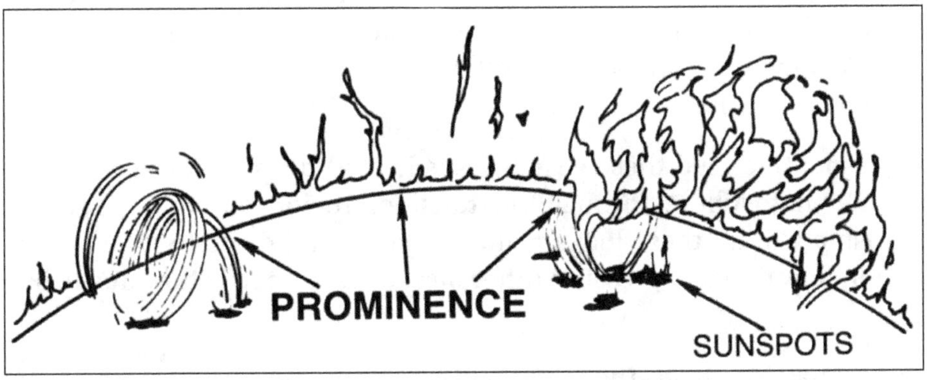

Illustration depicting solar prominences and sunspots.

Active regions involve all the phenomena directly linked to the magnetic field, which occur at different heights above the Sun's surface: sunspots and faculae, occur in the photosphere, spicules, Hα filaments and plages in the chromosphere, prominences in the chromosphere and transition region, and flares and coronal mass ejections happen in the corona and chromosphere. If flares are very violent, they can also perturb the photosphere and generate a Moreton wave. On the contrary, quiescent prominences are large, cool dense structures which are observed as dark, "snake-like" Hα ribbons (appearing like filaments) on the solar disc. Their temperature is about 5000–8000K, and so they are usually considered as chromospheric features.

In 2013, images from the High Resolution Coronal Imager revealed never-before-seen "magnetic braids" of plasma within the outer layers of these active regions.

Coronal Loops

TRACE 171Å coronal loops.

Coronal loops are the basic structures of the magnetic solar corona. These loops are the closed-magnetic flux cousins of the open-magnetic flux that can be found in coronal hole (polar) regions and the solar wind. Loops of magnetic flux well-up from the solar body and fill with hot solar plasma. Due to the heightened magnetic activity in these coronal loop regions, coronal loops can often be the precursor to solar flares and coronal mass ejections (CMEs).

The Solar plasma that feed these structures is heated from under 6000K to well over 106 K from the photosphere, through the transition region, and into the corona. Often, the solar plasma will fill these loops from one point and drain to another, called foot points (siphon flow due to a pressure difference, or asymmetric flow due to some other driver).

When the plasma rises from the foot points towards the loop top, as always occurs during the initial phase of a compact flare, it is defined as chromospheric evaporation. When the plasma rapidly cools and falls toward the photosphere, it is called chromospheric condensation. There may also be symmetric flow from both loop foot points, causing a build-up of mass in the loop structure. The plasma may cool rapidly in this region (for a thermal instability), its dark filaments obvious against the solar disk or prominences off the Sun's limb.

Coronal loops may have lifetimes in the order of seconds (in the case of flare events), minutes, hours or days. Where there is a balance in loop energy sources and sinks, coronal loops can last for long periods of time and are known as steady state or quiescent coronal loops.

Coronal loops are very important to our understanding of the current coronal heating problem. Coronal loops are highly radiating sources of plasma and are therefore easy to observe by instruments such as TRACE. An explanation of the coronal heating problem remains as these structures are being observed remotely, where many ambiguities are present (i.e. radiation contributions along the LOS). In-situ measurements are required before a definitive answer can be had, but due to the high plasma temperatures in the corona, in-situ measurements are, at present, impossible. The next mission of the NASA, the Parker Solar Probe will approach the Sun very closely allowing more direct observations.

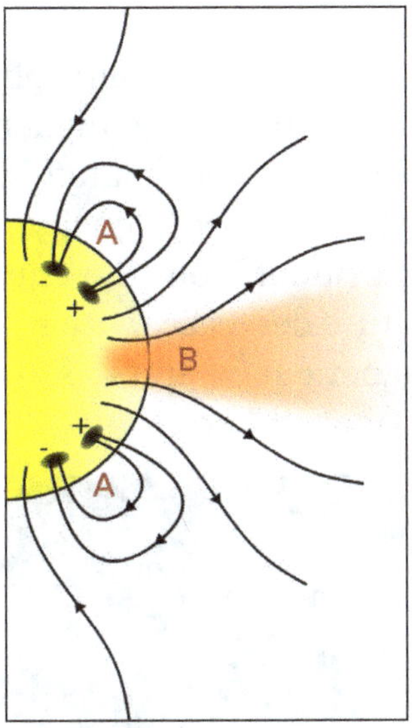

Coronal arches connecting regions of opposite magnetic polarity (A) and the unipolar magnetic field in the coronal hole (B).

Large-scale Structures

Large-scale structures are very long arcs which can cover over a quarter of the solar disk but contain plasma less dense than in the coronal loops of the active regions.

They were first detected in the June 8, 1968 flare observation during a rocket flight.

The large-scale structure of the corona changes over the 11-year solar cycle and becomes particularly simple during the minimum period, when the magnetic field of the Sun is almost similar to a dipolar configuration (plus a quadrupolar component).

Interconnections of Active Regions

The interconnections of active regions are arcs connecting zones of opposite magnetic field, of different active regions. Significant variations of these structures are often seen after a flare.

Some other features of this kind are helmet streamers—large cap-like coronal structures with long pointed peaks that usually overlie sunspots and active regions. Coronal streamers are considered to be sources of the slow solar wind.

Filament Cavities

Filament cavities are zones which look dark in the X-rays and are above the regions where Hα filaments are observed in the chromosphere. They were first observed in the two 1970 rocket flights which also detected coronal holes.

Filament cavities are cooler clouds of gases (plasma) suspended above the Sun's surface by magnetic forces. The regions of intense magnetic field look dark in images because they are empty of hot plasma. In fact, the sum of the magnetic pressure and plasma pressure must be constant everywhere on the heliosphere in order to have an equilibrium configuration: where the magnetic field is higher, the plasma must be cooler or less dense. The plasma pressure p can be calculated by the state equation of a perfect gas $p = nK_BT$, where n is the particle number density K_B, the Boltzmann constant and T the plasma temperature. It is evident from the equation that the plasma pressure lowers when the plasma temperature decreases with respect to the surrounding regions or when the zone of intense magnetic field empties. The same physical effect renders sunspots apparently dark in the photosphere.

A very long filament cavity is visible across the Sun's southern hemisphere.

Bright Points

Bright points are small active regions found on the solar disk. X-ray bright points were first detected on April 8, 1969 during a rocket flight.

The fraction of the solar surface covered by bright points varies with the solar cycle. They are associated with small bipolar regions of the magnetic field. Their average temperature ranges from (1.1E6 K) to (3.4E6 K). The variations in temperature are often correlated with changes in the X-ray emission.

Coronal Holes

Coronal holes are the Polar Regions which look dark in the X-rays since they do not emit much radiation. These are wide zones of the Sun where the magnetic field is unipolar and opens towards the interplanetary space. The high speed solar wind arises mainly from these regions.

In the UV images of the coronal holes, some small structures, similar to elongated bubbles, are often seen as they were suspended in the solar wind. These are the coronal plumes. More exactly, they are long thin streamers that project outward from the Sun's north and south poles.

The Quiet Sun

The solar regions which are not part of active regions and coronal holes are commonly identified as the quiet Sun.

The equatorial region has a faster rotation speed than the polar zones. The result of the Sun's differential rotation is that the active regions always arise in two bands parallel to the equator and their extension increases during the periods of maximum of the solar cycle, while they almost disappear during each minimum. Therefore, the quiet Sun always coincides with the equatorial zone and its surface is less active during the maximum of the solar cycle. Approaching the minimum of the solar cycle (also named butterfly cycle), the extension of the quiet Sun increases until it covers the whole disk surface excluding some bright points on the hemisphere and the poles, where there are coronal holes.

Variability of the Corona

A portrait as diversified as the one already pointed out for the coronal features is emphasized by the analysis of the dynamics of the main structures of the corona, which evolve in times very different among them. Studying the coronal variability in its complexity is not easy because the times of evolution of the different structures can vary considerably: from seconds to several months. The typical sizes of the regions where coronal events take place vary in the same way, as it is shown in the following table.

Coronal event	Typical time-scale	Typical length-scale (Mm)
Active region flare	10 to 10,000 seconds	10–100
X-ray bright point	minutes	1–10
Transient in large-scale structures	from minutes to hours	~100
Transient in interconnecting arcs	from minutes to hours	~100
Quiet Sun	from hours to months	100–1000
Coronal hole	several rotations	100–1000

Flares

On August 31, 2012 a long filament of solar material that had been hovering in the Sun's outer atmosphere, the corona.

Flares take place in active regions and are characterized by a sudden increase of the radiative flux emitted from small regions of the corona. They are very complex phenomena, visible at different wavelengths; they involve several zones of the solar atmosphere and many physical effects, thermal and not thermal, and sometimes wide reconnections of the magnetic field lines with material expulsion.

Flares are impulsive phenomena, of average duration of 15 minutes, and the most energetic events can last several hours. Flares produce a high and rapid increase of the density and temperature.

An emission in white light is only seldom observed: usually, flares are only seen at extreme UV wavelengths and into the X-rays, typical of the chromospheric and coronal emission.

In the corona, the morphology of flares, is described by observations in the UV, soft and hard X-rays, and in Hα wavelengths, and is very complex. However, two kinds of basic structures can be distinguished:

- Compact flares, when each of the two arches where the event is happening maintains its morphology: only an increase of the emission is observed without significant structural variations. The emitted energy is of the order of $10^{22} - 10^{23}$ J.
- Flares of long duration, associated with eruptions of prominences, transients in white light and two-ribbon flares: in this case the magnetic loops change their configuration during the event. The energies emitted during these flares are of such great proportion they can reach 10^{25} J.

As for temporal dynamics, three different phases are generally distinguished, whose duration are not comparable. The durations of those periods depend on the range of wavelengths used to observe the event:

- An initial impulsive phase, whose duration is on the order of minutes, strong emissions of energy are often observed even in the microwaves, EUV wavelengths and in the hard X-ray frequencies.

- A maximum phase.
- A decay phase, which can last several hours.

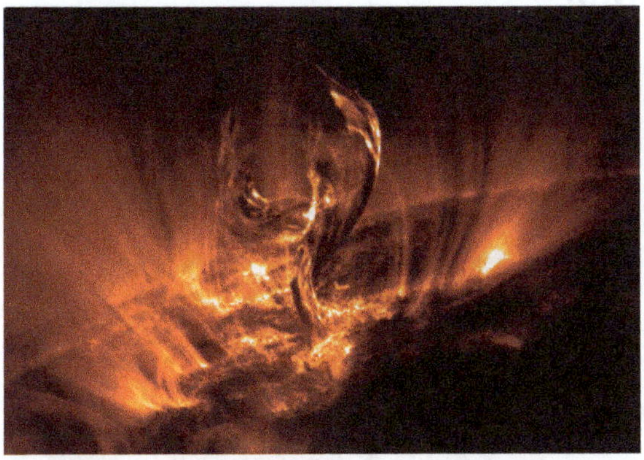

Filament erupting during a solar flare, seen at EUV wavelengths (TRACE).

Sometimes also a phase preceding the flare can be observed, usually called as "preflare" phase.

Transients

Accompanying solar flares or large solar prominences, "coronal transients" (also called coronal mass ejections) are sometimes released. These are enormous loops of coronal material that travel outward from the Sun at over a million kilometers per hour, containing roughly 10 times the energy of the solar flare or prominence that accompanies them. Some larger ejections can propel hundreds of millions of tons of material into space at roughly 1.5 million kilometers an hour.

Stellar Coronae

Coronal stars are ubiquitous among the stars in the cool half of the Hertzsprung–Russell diagram. These coronae can be detected using X-ray telescopes. Some stellar coronae, particularly in young stars, are much more luminous than the Sun's. For example, FK Comae Berenices is the prototype for the FK Com class of variable star. These are giants of spectral types G and K with an unusually rapid rotation and signs of extreme activity. Their X-ray coronae are among the most luminous ($L_x \geq 10^{32}$ erg·s^{-1} or 10^{25}W) and the hottest known with dominant temperatures up to 40 MK.

The astronomical observations planned with the Einstein Observatory by Giuseppe Vaiana and his group showed that F-, G-, K- and M-stars have chromospheres and often coronae much like our Sun. The O-B stars, which do not have surface convection zones, have a strong X-ray emission. However these stars do not have coronae, but the outer stellar envelopes emit this radiation during shocks due to thermal instabilities in rapidly moving gas blobs. Also A-stars do not have convection zones but they do not emit at the UV and X-ray wavelengths. Thus they appear to have neither chromospheres nor coronae.

Physics of the Corona

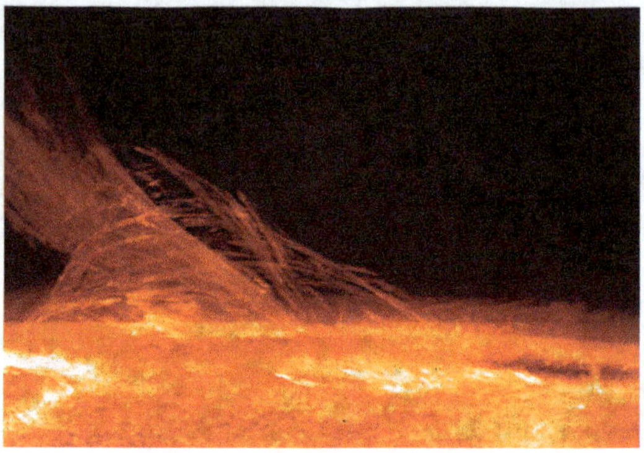

The matter in the external part of the solar atmosphere is in the state of plasma, at very high temperature (a few million kelvin) and at very low density (of the order of 1015 particles/m^3). According to the definition of plasma, it is a quasi-neutral ensemble of particles which exhibits a collective behaviour.

The composition is similar to that in the Sun's interior, mainly hydrogen, but with much greater ionization than that found in the photosphere. Heavier metals, such as iron, are partially ionized and have lost most of the external electrons. The ionization state of a chemical element depends strictly on the temperature and is regulated by the Saha equation in the lowest atmosphere, but by collisional equilibrium in the optically-thin corona. Historically, the presence of the spectral lines emitted from highly ionized states of iron allowed determination of the high temperature of the coronal plasma, revealing that the corona is much hotter than the internal layers of the chromosphere.

The corona behaves like a gas which is very hot but very light at the same time: the pressure in the corona is usually only 0.1 to 0.6 Pa in active regions, while on the Earth the atmospheric pressure is about 100 kPa, approximately a million times higher than on the solar surface. However it is not properly a gas, because it is made of charged particles, basically protons and electrons, moving at different velocities. Supposing that they have the same kinetic energy on average (for the equipartition theorem), electrons have a mass roughly 1800 times smaller than protons, therefore they acquire more velocity. Metal ions are always slower. This fact has relevant physical consequences either on radiative processes (that are very different from the photospheric radiative processes), or on thermal conduction. Furthermore, the presence of electric charges induces the generation of electric currents and high magnetic fields. Magnetohydrodynamic waves (MHD waves) can also propagate in this plasma, even if it is not still clear how they can be transmitted or generated in the corona.

Radiation

The corona emits radiation mainly in the X-rays, observable only from space. The plasma is transparent to its own radiation and to that one coming from below, therefore we say that it is optically-thin. The gas, in fact, is very rarefied and the photon mean free-path overcomes by far all the other length-scales, including the typical sizes of the coronal features.

Different processes of radiation take place in the emission, due to binary collisions between plasma particles, while the interactions with the photons, coming from below; are very rare. Because the emission is due to collisions between ions and electrons, the energy emitted from a unit volume in the time unit is proportional to the squared number of particles in a unit volume, or more exactly, to the product of the electron density and proton density.

Thermal Conduction

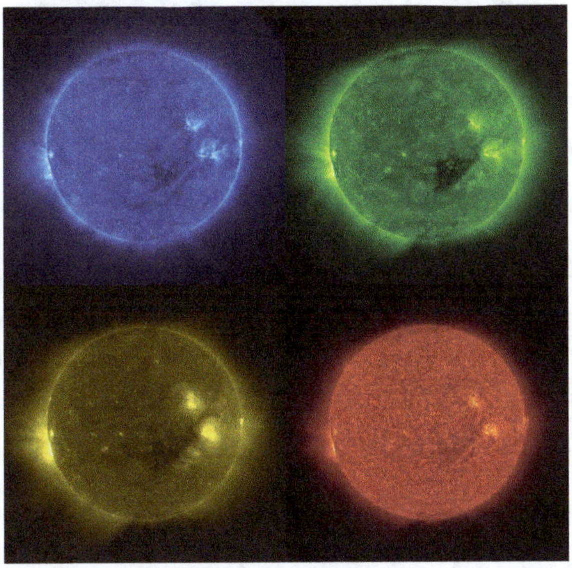

A mosaic of the extreme ultraviolet images taken from STEREO on December 4, 2006. These false color images show the Sun's atmospheres at a range of different temperatures. Clockwise from top left: 1 million degrees C (171 Å—blue), 1.5 million °C (195Å—green), 60000–80000°C (304 Å—red), and 2.5 million °C (286 Å—yellow).

In the corona thermal conduction occurs from the external hotter atmosphere towards the inner cooler layers. Responsible for the diffusion process of the heat are the electrons, which are much lighter than ions and move faster.

When there is a magnetic field the thermal conductivity of the plasma becomes higher in the direction which is parallel to the field lines rather than in the perpendicular direction. A charged particle moving in the direction perpendicular to the magnetic field line is subject to the Lorentz force which is normal to the plane individuated by the velocity and the magnetic field. This force bends the path of the particle. In general, since particles also have a velocity component along the magnetic field line, the Lorentz force constrains them to bend and move along spirals around the field lines at the cyclotron frequency.

STEREO – First images as a slow animation.

If collisions between the particles are very frequent, they are scattered in every direction. This happens in the photosphere, where the plasma carries the magnetic field in its motion. In the corona, on the contrary, the mean free-path of the electrons is of the order of kilometres and even more, so each electron can do a helicoidal motion long before being scattered after a collision. Therefore, the heat transfer is enhanced along the magnetic field lines and inhibited in the perpendicular direction.

In the direction longitudinal to the magnetic field, the thermal conductivity of the corona is;

$$k = 20\left(\frac{2}{\pi}\right)^{3/2} \frac{(k_B T)^{5/2} k_B}{m_e^{1/2} e^4 \ln \Lambda} \approx 1.810^{-10} \frac{T^{5/2}}{\ln \Lambda} W m^{-1} K^{-1}$$

where k_B is the Boltzmann constant, T is the temperature in kelvin, m_e the electron mass, e the electric charge of the electron,

$$\ln \Lambda = \ln\left(12\pi n \lambda_D^3\right)$$

the Coulomb logarithm, and

$$\lambda_D = \sqrt{\frac{k_B T}{4\pi n e^2}}$$

the Debye length of the plasma with particle density n. The Coulomb logarithm $\ln\Lambda$ is roughly 20 in the corona, with a mean temperature of 1 MK and a density of 10^{15} particles/m³, and about 10 in the chromosphere, where the temperature is approximately 10kK and the particle density is of the order of 10^{18} particles/m³, and in practice it can be assumed constant.

Thence, if we indicate with q the heat for a volume unit, expressed in J m^{-3}, the Fourier equation of heat transfer, to be computed only along the direction x of the field line, becomes:

$$\frac{\partial q}{\partial t} = 0.910^{-11}\frac{\partial^2 T^{7/2}}{\partial x^2}.$$

Numerical calculations have shown that the thermal conductivity of the corona is comparable to that of copper.

Coronal Seismology

Coronal seismology is a new way of studying the plasma of the solar corona with the use of magnetohydrodynamic (MHD) waves. Magnetohydrodynamics studies the dynamics of electrically conducting fluids—in this case the fluid is the coronal plasma. Philosophically, coronal seismology is similar to the Earth's seismology, the Sun's helioseismology, and MHD spectroscopy of laboratory plasma devices. In all these approaches, waves of various kinds are used to probe a medium. The potential of coronal seismology in the estimation of the coronal magnetic field, density scale height, fine structure and heating has been demonstrated by different research groups.

Coronal Heating Problem

The coronal heating problem in solar physics relates to the question of why the temperature of the Sun's corona is millions of kelvin higher than that of the surface. Several theories have been proposed to explain this phenomenon but it is still challenging to determine which of these is correct. The problem first emerged when Bengt Edlen and Walter Grotrian identified Fe IX and Ca XIV lines in the solar spectrum. This led to the discovery that the emission lines seen during solar eclipses are not caused by an unknown element called "coronium" but known elements at very high stages of ionization. The comparison of the coronal and the photospheric temperatures of 6000K, leads to the question of how the 200 times hotter coronal temperature can be maintained. The problem is primarily concerned with how the energy is transported up into the corona and then convert it into heat within a few solar radii.

The high temperatures require energy to be carried from the solar interior to the corona by non-thermal processes, because the second law of thermodynamics prevents heat from flowing directly from the solar photosphere (surface), which is at about 5800K, to the much hotter corona at about 1 to 3 MK (parts of the corona can even reach 10MK).

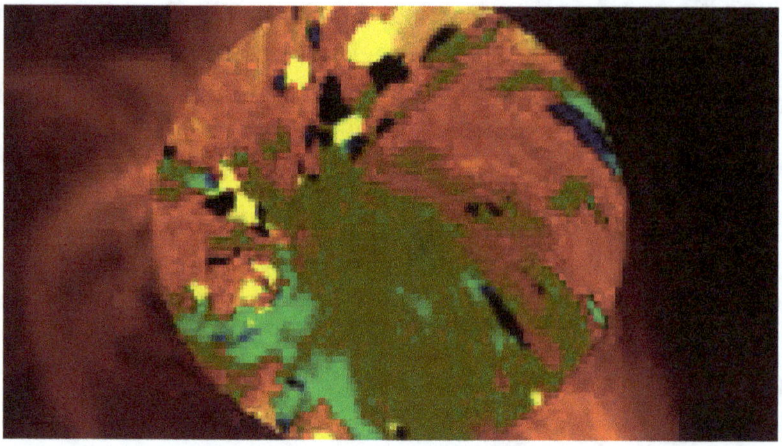

A new visualisation technique can provide clues to the coronal heating problem.

Between the photosphere and the corona, the thin region through which the temperature increases is known as the transition region. It ranges from only tens to hundreds of kilometers thick. Energy cannot be transferred from the cooler photosphere to the corona by conventional heat transfer as this would violate the second law of thermodynamics. An analogy of this would be a light bulb raising the temperature of the air surrounding it to something greater than its glass surface. Hence, some other manner of energy transfer must be involved in the heating of the corona.

The amount of power required to heat the solar corona can easily be calculated as the difference between coronal radiative losses and heating by thermal conduction toward the chromosphere through the transition region. It is about 1 kilowatt for every square meter of surface area on the Sun's chromosphere, or 1/40000 of the amount of light energy that escapes the Sun.

Many coronal heating theories have been proposed, but two theories have remained as the most likely candidates: wave heating and magnetic reconnection (or nanoflares). Through most of the past 50 years, neither theory has been able to account for the extreme coronal temperatures.

In 2012, high resolution (<0.2") soft X-ray imaging with the High Resolution Coronal Imager aboard a sounding rocket revealed tightly wound braids in the corona. It is hypothesized that the reconnection and unravelling of braids can act as primary sources of heating of the active solar corona to temperatures of up to 4 million kelvin. The main heat source in the quiescent corona (about 1.5 million kelvin) is assumed to originate from MHD waves.

The NASA mission Parker Solar Probe is intended to approach the Sun to a distance of approximately 9.5 solar radii to investigate coronal heating and the origin of the solar wind. It has been successfully launched on August 12, 2018 and is currently leaving Earth for the first orbit.

<table>
<tr><td colspan="3">Competing heating mechanisms</td></tr>
<tr><td colspan="3">Heating Models</td></tr>
<tr><td>Hydrodynamic</td><td colspan="2">Magnetic</td></tr>
<tr><td rowspan="2">• No magnetic field
• Slow rotating stars</td><td>DC (reconnection)</td><td>AC (waves)</td></tr>
<tr><td>• B-field stresses
• Reconnection events
• Flares-nanoflares
• Uniform heating rates</td><td>• Photospheric foot point shuffling
• MHD wave propagation
• High Alfvén wave flux
• Non-uniform heating rates</td></tr>
</table>

Wave Heating Theory

The wave heating theory, proposed in 1949 by Evry Schatzman, proposes that waves carry energy from the solar interior to the solar chromosphere and corona. The Sun is made of plasma rather than ordinary gas, so it supports several types of waves analogous to sound waves in air. The most important types of wave are magneto-acoustic waves and Alfvén waves. Magneto-acoustic waves are sound waves that have been modified by the presence of a magnetic field, and Alfvén waves are similar to ultra low frequency radio waves that have been modified by interaction with matter in the plasma. Both types of waves can be launched by the turbulence of granulation and super granulation at the solar photosphere, and both types of waves can carry energy for some distance through the solar atmosphere before turning into shock waves that dissipate their energy as heat.

One problem with wave heating is delivery of the heat to the appropriate place. Magneto-acoustic waves cannot carry sufficient energy upward through the chromosphere to the corona, both because of the low pressure present in the chromosphere and because they tend to be reflected back to the photosphere. Alfvén waves can carry enough energy, but do not dissipate that energy rapidly enough once they enter the corona. Waves in plasmas are notoriously difficult to understand and describe analytically, but computer simulations, carried out by Thomas Bogdan and colleagues in 2003, seem to show that Alfvén waves can transmute into other wave modes at the base of the corona, providing a pathway that can carry large amounts of energy from the photosphere through the chromosphere and transition region and finally into the corona where it dissipates it as heat.

Another problem with wave heating has been the complete absence, until the late 1990s, of any direct evidence of waves propagating through the solar corona. The first direct observation of waves propagating into and through the solar corona was made in 1997 with the Solar and Heliospheric Observatory space-borne solar observatory, the first platform capable of observing the Sun in the extreme ultraviolet (EUV) for long periods of time with stable photometry. Those were magneto-acoustic waves with a frequency of about 1 millihertz (mHz, corresponding to a 1000 second wave period), that carry only about 10% of the energy required to heat the corona. Many observations exist of localized wave phenomena, such as Alfvén waves launched by solar flares, but those events are transient and cannot explain the uniform coronal heat.

It is not yet known exactly how much wave energy is available to heat the corona. Results published in 2004 using data from the TRACE spacecraft seem to indicate that there are waves in the solar atmosphere at frequencies as high as 100mHz (10 second period). Measurements of the temperature of different ions in the solar wind with the UVCS instrument aboard SOHO give strong indirect evidence that there are waves at frequencies as high as 200Hz, well into the range of human hearing. These waves are very difficult to detect under normal circumstances, but evidence collected during solar eclipses by teams from Williams College suggest the presences of such waves in the 1^{-10} Hz range.

Recently, Alfvénic motions have been found in the lower solar atmosphere and also in the quiet Sun, in coronal holes and in active regions using observations with AIA on board the Solar Dynamics Observatory. These Alfvénic oscillations have significant power, and seem to be connected to the chromospheric Alfvénic oscillations previously reported with the Hinode spacecraft.

Solar wind observations with the Wind spacecraft have recently shown evidence to support theories of Alfvén-cyclotron dissipation, leading to local ion heating.

Magnetic Reconnection Theory

The magnetic reconnection theory relies on the solar magnetic field to induce electric currents in the solar corona. The currents then collapse suddenly, releasing energy as heat and wave energy in the corona. This process is called "reconnection" because of the peculiar way that magnetic fields behave in plasma (or any electrically conductive fluid such as mercury or seawater). In a plasma, magnetic field lines are normally tied to individual pieces of matter, so that the topology of the magnetic field remains the same: if a particular north and south magnetic pole are connected by a single field line, then even if the plasma is stirred or if the magnets are moved around, that field line will continue to connect those particular poles. The connection is maintained by electric currents that are induced in the plasma. Under certain conditions, the electric currents can collapse, allowing the magnetic field to "reconnect" to other magnetic poles and release heat and wave energy in the process.

Magnetic reconnection is hypothesized to be the mechanism behind solar flares, the largest explosions in the Solar System. Furthermore, the surface of the Sun is covered with millions of small magnetized regions 50–1000km across. These small magnetic poles are buffeted and churned by the constant granulation. The magnetic field in the solar corona must undergo nearly constant reconnection to match the motion of this "magnetic carpet", so the energy released by the reconnection is a natural candidate for the coronal heat, perhaps as a series of "microflares" that individually provide very little energy but together account for the required energy.

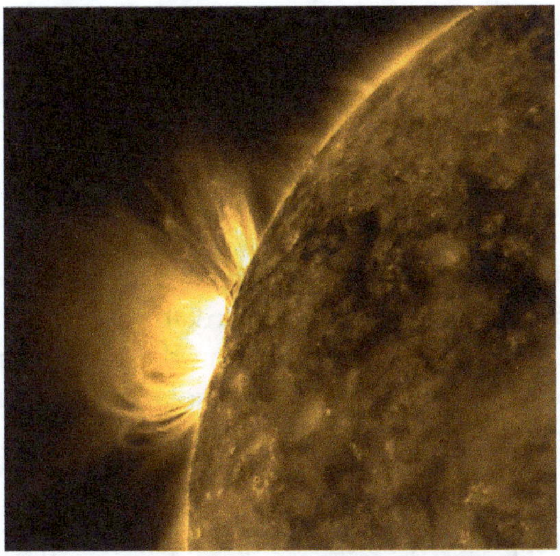

Arcing active region by Solar Dynamics Observatory.

The idea that nanoflares might heat the corona was proposed by Eugene Parker in the 1980s but is still controversial. In particular, ultraviolet telescopes such as TRACE and SOHO/EIT can observe individual micro-flares as small brightenings in extreme ultraviolet light, but there seem to be too few of these small events to account for the energy released into the corona. The additional energy not accounted for could be made up by wave energy, or by gradual magnetic reconnection that releases energy more smoothly than micro-flares and therefore doesn't appear well in the TRACE data. Variations on the micro-flare hypothesis use other mechanisms to stress the magnetic field or to release the energy, and are a subject of active research in 2005.

Spicules (Type II)

For decades, researchers believed spicules could send heat into the corona. However, following observational research in the 1980s, it was found that spicule plasma did not reach coronal temperatures, and so the theory was discounted.

As per studies performed in 2010, a new class of spicules (TYPE II) discovered in 2007, which travel faster (up to 100 km/s) and have shorter lifespans, can account for the problem. These jets insert heated plasma into the Sun's outer atmosphere.

Thus, a much greater understanding of the Corona and improvement in the knowledge of the Sun's subtle influence on the Earth's upper atmosphere can be expected henceforth. The Atmospheric Imaging Assembly on NASA's recently launched Solar Dynamics Observatory and NASA's Focal Plane Package for the Solar Optical Telescope on the Japanese Hinode satellite which was used to test this hypothesis. The high spatial and temporal resolutions of the newer instruments reveal this coronal mass supply.

These observations reveal a one-to-one connection between plasma that is heated to millions of degrees and the spicules that insert this plasma into the corona.

Coronal Mass Ejection

A coronal mass ejection (CME) is a significant release of plasma and accompanying magnetic field from the solar corona. They often follow solar flares and are normally present during a solar prominence eruption. The plasma is released into the solar wind, and can be observed in coronagraph imagery.

Coronal mass ejections are often associated with other forms of solar activity, but a broadly accepted theoretical understanding of these relationships has not been established. CMEs most often originate from active regions on the Sun's surface, such as groupings of sunspots associated with frequent flares. Near solar maxima, the Sun produces about three CMEs every day, whereas near solar minima, there is about one CME every five days.

The largest recorded geomagnetic perturbation, resulting presumably from a CME hitting the Earth's magnetosphere, was the solar storm of 1859 (the Carrington Event), which took down parts of the recently created US telegraph network, starting fires and shocking some telegraph operators.

Coronal mass ejections release large quantities of matter and electromagnetic radiation into space above the Sun's surface, either near the corona (sometimes called a solar prominence), or farther into the planetary system, or beyond (interplanetary CME). The ejected material is a magnetized plasma consisting primarily of electrons and protons. While solar flares are very fast (being electromagnetic radiation), CMEs are relatively slow.

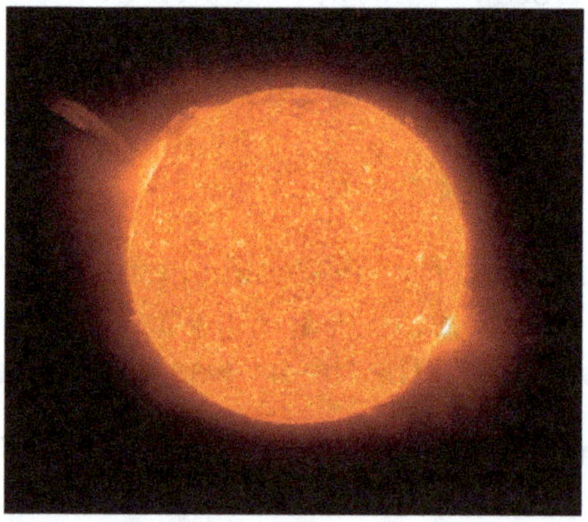

Arcs rise above an active region on the surface of the Sun.

Coronal mass ejections are associated with enormous changes and disturbances in the coronal magnetic field. They are usually observed with a white-light coronagraph.

Cause

The phenomenon of magnetic reconnection is closely associated with CMEs and solar flares. In magnetohydrodynamic theory, the sudden rearrangement of magnetic field lines when two oppositely directed magnetic fields are brought together is called "magnetic reconnection". Reconnection releases energy stored in the original stressed magnetic fields. These magnetic field lines can become twisted in a helical structure, with a 'right-hand twist' or a 'left hand twist'. As the Sun's magnetic field lines become more and more twisted, CMEs appear to be a 'valve' to release the magnetic energy being built up, as evidenced by the helical structure of CMEs, that would otherwise renew itself continuously each solar cycle and eventually rip the Sun apart.

On the Sun, magnetic reconnection may happen on solar arcades—a series of closely occurring loops of magnetic lines of force. These lines of force quickly reconnect into a low arcade of loops, leaving a helix of magnetic field unconnected to the rest of the arcade. The sudden release of energy during this process causes the solar flare and ejects the CME. The helical magnetic field and the material that it contains may violently expand outwards forming a CME. This also explains why CMEs and solar flares typically erupt from what are known as the active regions on the Sun where magnetic fields are much stronger on average.

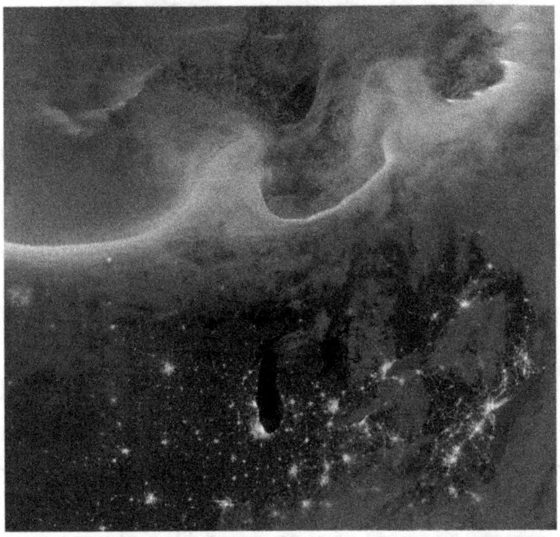

Satellite photo of aurora borealis stretching across Quebec and Ontario in the early morning.

Impact on Earth

When the ejection is directed towards Earth and reaches it as an interplanetary CME (ICME), the shock wave of traveling mass causes a geomagnetic storm that may disrupt Earth's magnetosphere, compressing it on the day side and extending the night-side magnetic tail. When the magnetosphere reconnects on the nightside, it releases power on the order of terawatt scale, which is directed back toward Earth's upper atmosphere.

Solar energetic particles can cause particularly strong aurorae in large regions around Earth's magnetic poles. These are also known as the *Northern Lights* in the northern hemisphere, and the *Southern Lights* in the southern hemisphere. Coronal mass ejections, along with solar flares of other origin, can disrupt radio transmissions and cause damage to satellites and electrical transmission line facilities, resulting in potentially massive and long-lasting power outages.

Energetic protons released by a CME can cause an increase in the number of free electrons in the ionosphere, especially in the high-latitude polar regions. The increase in free electrons can enhance radio wave absorption, especially within the D-region of the ionosphere, leading to Polar Cap Absorption (PCA) events.

Humans at high altitudes, as in airplanes or space stations, risk exposure to relatively intense solar particle events. The energy absorbed by astronauts is not reduced by a typical spacecraft shield design and, if any protection is provided, it would result from changes in the microscopic inhomogeneity of the energy absorption events.

Physical Properties

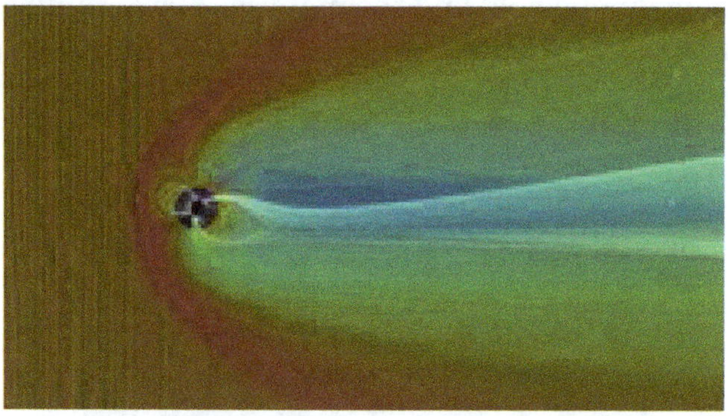

This video features two model runs. One looks at a moderate coronal mass ejection (CME) from 2006. The second run examines the consequences of a large coronal mass ejection, such as the Carrington-class CME of 1859.

A typical coronal mass ejection may have any or all of three distinctive features: a cavity of low electron density, a dense core (the prominence, which appears on coronagraph images as a bright region embedded in this cavity), and a bright leading edge.

Most ejections originate from active regions on the Sun's surface, such as groupings of sunspots associated with frequent flares. These regions have closed magnetic field lines, in which the magnetic field strength is large enough to contain the plasma. These field lines must be broken or weakened for the ejection to escape from the Sun. However, CMEs may also be initiated in quiet surface regions, although in many cases the quiet region was recently active. During solar minimum, CMEs form primarily in the coronal streamer belt near the solar magnetic equator. During solar maximum, they originate from active regions whose latitudinal distribution is more homogeneous.

Coronal mass ejections reach velocities from 20 to 3,200 km/s (12 to 1,988 mi/s) with an average speed of 489 km/s (304 mi/s), based on SOHO/LASCO measurements between 1996 and 2003. These speeds correspond to transit times from the Sun out to the mean radius of Earth's orbit of about 13 hours to 86 days (extremes), with about 3.5 days as the average. The average mass ejected is 1.6×10^{12} kg (3.5×10^{12} lb). However, the estimated mass values for CMEs are only lower limits, because coronagraph measurements provide only two-dimensional data. The frequency of ejections depends on the phase of the solar cycle: from about 0.2 per day near the solar minimum to 3.5 per day near the solar maximum. These values are also lower limits because ejections propagating away from Earth (backside CMEs) usually cannot be detected by coronagraphs.

Current knowledge of coronal mass ejection kinematics indicates that the ejection starts with an initial pre-acceleration phase characterized by a slow rising motion, followed by a period of rapid acceleration away from the Sun until a near-constant velocity is reached. Some balloon CMEs, usually the slowest ones, lack this three-stage evolution, instead accelerating slowly and continuously throughout their flight. Even for CMEs with a well-defined acceleration stage, the pre-acceleration stage is often absent, or perhaps unobservable.

Association with other Solar Phenomena

Coronal mass ejections are often associated with other forms of solar activity, most notably:

- Solar flares.
- Eruptive prominence and X-ray sigmoids.
- Coronal dimming (long-term brightness decrease on the solar surface).
- Moreton waves.
- Coronal waves (bright fronts propagating from the location of the eruption).
- Post-eruptive arcades.

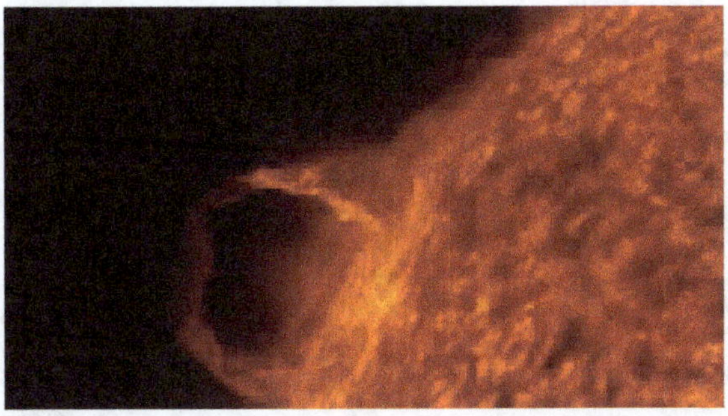

The association of a CME with some of those phenomena is common but not fully understood. For example, CMEs and flares are normally closely related, but there was confusion about this point caused by the events originating beyond the limb. For such events no flare could be detected. Most weak flares do not have associated CMEs; most powerful ones do. Some CMEs occur without any flare-like manifestation, but these are the weaker and slower ones. It is now thought that CMEs and associated flares are caused by a common event (the CME peak acceleration and the flare impulsive phase generally coincide). In general, all of these events (including the CME) are thought to be the result of a large-scale restructuring of the magnetic field; the presence or absence of a CME during one of these restructures would reflect the coronal environment of the process (i.e., can the eruption be confined by overlying magnetic structure, or will it simply break through and enter the solar wind).

Theoretical Models

It was first postulated that CMEs might be driven by the heat of an explosive flare. However, it soon became apparent that many CMEs were not associated with flares, and that even those that were often started before the flare. Because CMEs are initiated in the solar corona (which is dominated by magnetic energy), their energy source must be magnetic.

Because the energy of CMEs is so high, it is unlikely that their energy could be directly driven by emerging magnetic fields in the photosphere (although this is still a possibility). Therefore, most models of CMEs assume that the energy is stored up in the coronal magnetic field over a long period of time and then suddenly released by some instability or a loss of equilibrium in the field. There is still no consensus on which of these release mechanisms is correct, and observations are not currently able to constrain these models very well. These same considerations apply equally well to solar flares, but the observable signatures of these phenomena differ.

Interplanetary Coronal Mass Ejections

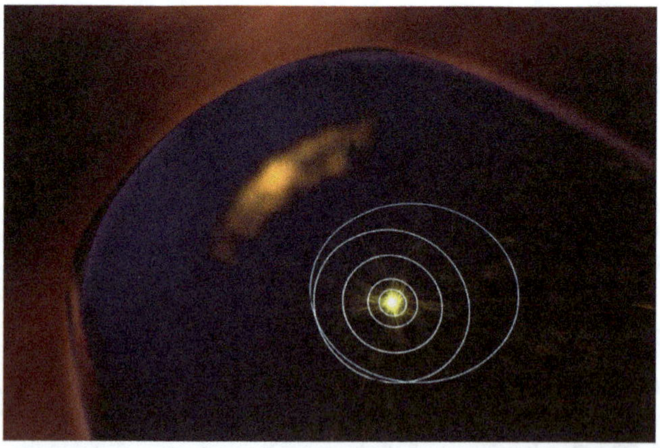

Illustration of a coronal mass ejection moving beyond the planets toward the heliopause.

CMEs typically reach Earth one to five days after leaving the Sun. During their propagation, CMEs interact with the solar wind and the interplanetary magnetic field (IMF). As a consequence, slow CMEs are accelerated toward the speed of the solar wind and fast CMEs are decelerated toward the speed of the solar wind. The strongest deceleration or acceleration occurs close to the Sun, but it can continue even beyond Earth orbit (1 AU), which was observed using measurements at Mars and by the Ulysses spacecraft. CMEs faster than about 500 km/s (310 mi/s) eventually drive a shock wave. This happens when the speed of the CME in the frame of reference moving with the solar wind is faster than the local fast magnetosonic speed. Such shocks have been observed directly by coronagraphs in the corona, and are related to type II radio bursts. They are thought to form sometimes as low as 2 $R_{\odot}$ (solar radii). They are also closely linked with the acceleration of solar energetic particles.

Future Risk

According to a report published in 2012 by physicist Pete Riley of Predictive Science Inc., the chance of Earth being hit by a Carrington-class storm between 2012 and 2022 is 12%.

Stellar Coronal Mass Ejections

There have been a small number of CMEs observed on other stars, all of which as of 2016 have been found on M dwarfs. These have been detected by spectroscopy, most often by studying Balmer lines: the material ejected toward the observer causes asymmetry in the blue wing of the line profiles due to Doppler shift. This enhancement can be seen in absorption when it occurs on the stellar disc (the material is cooler than its surrounding), and in emission when it is outside the disc. The observed projected velocities of CMEs range from ≈84 to 5,800 km/s (52 to 3,600 mi/s). Compared to activity on the Sun, CME activity on other stars seems to be far less common.

Coronal Loop

Coronal loops are huge loops of magnetic field beginning and ending on the Sun's visible surface (photosphere) projecting into the solar atmosphere (corona). Hot glowing ionized gas (plasma) trapped in the loops makes them visible. Coronal loops range widely in size up to several thousand kilometers long. They are transient features of the solar surface, forming and dissipating over periods of seconds to days. They form the basic structure of the lower corona and transition region of the Sun. These highly structured loops are a direct consequence of the twisted solar magnetic flux within the solar body. Coronal loops are associated with sunspots; the two "footpoints" where the loop passes through the sun's surface are often sunspots. This is because sunspots occur at regions of high magnetic field. The high magnetic field where the loop passes through the surface forms a barrier to convection currents, which bring hot plasma from the interior to the sun's surface, so the plasma in these high field regions is cooler than the rest of the sun's surface, appearing as a dark spot when viewed against the rest of the photosphere. The population of coronal loops varies with the 11 year solar cycle, which also influences the number of sunspots.

Physical Features

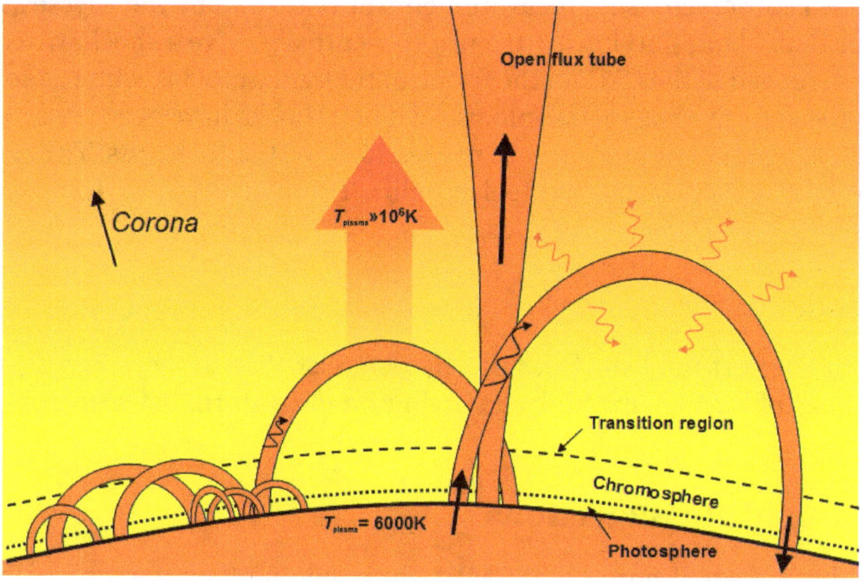

Diagram of the low corona and transition region, where many scales of coronal loops can be observed.

Due to a natural process called the solar dynamo driven by heat produced in the Sun's core, motion of the electrically conductive ionized gas (plasma) which makes up the Sun creates electric currents, which in turn create powerful magnetic fields in the Sun's interior. These magnetic fields are in the form of closed loops of magnetic flux, which are twisted and tangled by the different rotation rates of the gas at different latitudes of the solar sphere. A coronal loop occurs when a curved arc of the magnetic field projects through the visible surface of the Sun, the photosphere, protruding into the solar atmosphere.

Within the magnetic field, the paths of the moving electrically charged particles (electrons and ions) which make up the Sun's gas are sharply bent by the field when moving transverse to the field, so they can only move freely parallel to the magnetic field lines, tending to spiral around the lines. Thus the gas within a coronal loop cannot escape sideways out of the loop but only flow along its length. The higher temperature in the Sun's atmosphere causes this gas to glow, making the loop visible through telescopes. Coronal loops are ideal structures to observe when trying to understand the transfer of energy from the solar body, through the transition region and into the corona.

The strong interaction of the magnetic field with the dense plasma on and below the sun's surface tends to cause the magnetic field lines to be "tied" to the motion of the sun's gas, so the two 'footpoints' where the loop enters the photosphere are fixed to the sun's surface, and rotate with the surface. Within each footpoint, which may be up to 100,000 km in diameter, the strong magnetic flux tends to inhibit the convection currents which carry hot gas from the sun's interior to the surface, so the footpoints are often (but not always) cooler than the surrounding photosphere. These appear as dark spots on the sun's surface; sunspots. Thus sunspots tend to come in pairs of opposite magnetic polarity; one a point where the magnetic field loop emerges from the photosphere, and the other where the loop enters the surface again. The footpoint where the magnetic field lines emerge from the surface is thus a North magnetic pole while the point where the field lines enter is a South magnetic pole.

Coronal loops form in a wide range of sizes, from 10 km to 10,000 km. A related phenomenon, open flux tubes of magnetic field extend from the surface far into the corona and heliosphere and are the source of the sun's large scale magnetic field (magnetosphere) and the solar wind. Coronal loops have a wide variety of temperatures along their lengths. Loops at temperatures below 1 megakelvin (MK) are generally known as cool loops, those existing at around 1 MK are known as warm loops, and those beyond 1 MK are known as hot loops. Naturally, these different categories radiate at different wavelengths.

Location

Coronal loops populate both active and quiet regions of the solar surface. Active regions on the solar surface take up small areas but produce the majority of activity and are often the source of flares and Coronal Mass Ejections due to the intense magnetic field present. Active regions produce 82% of the total coronal heating energy. Coronal holes are open field lines located predominantly in the polar regions of the Sun and are known to be the source of the fast solar wind. The quiet Sun makes up the rest of the solar surface. The quiet Sun, although less active than active regions, is awash with dynamic processes and transient events (bright points, nanoflares and jets). As a general rule, the quiet Sun exists in regions of closed magnetic structures, and active regions are highly dynamic sources of explosive events. It is important to note that observations suggest the whole corona is massively populated by open and closed magnetic fieldlines.

Coronal Loops and the Coronal Heating Problem

A closed fieldline does not constitute a coronal loop; however, closed flux must be *filled with plasma* before it can be called a coronal loop. With this in mind, it becomes clear that coronal loops are a rarity on the solar surface, as the majority of closed-flux structures are empty. This means the mechanism that heats the corona and injects chromospheric plasma into the closed magnetic flux is highly localised. The mechanism behind plasma filling, dynamic flows and coronal heating remains a mystery. The mechanism(s) must be stable enough to continue to feed the corona with chromospheric plasma and powerful enough to accelerate and therefore heat the plasma from 6000 K to well over 1 MK over the short distance from the chromosphere and transition region to the corona. This is the very reason coronal loops are targeted for intense study. They are anchored to the photosphere, are fed by chromospheric plasma, protrude into the transition region and exist at coronal temperatures after undergoing intensive heating.

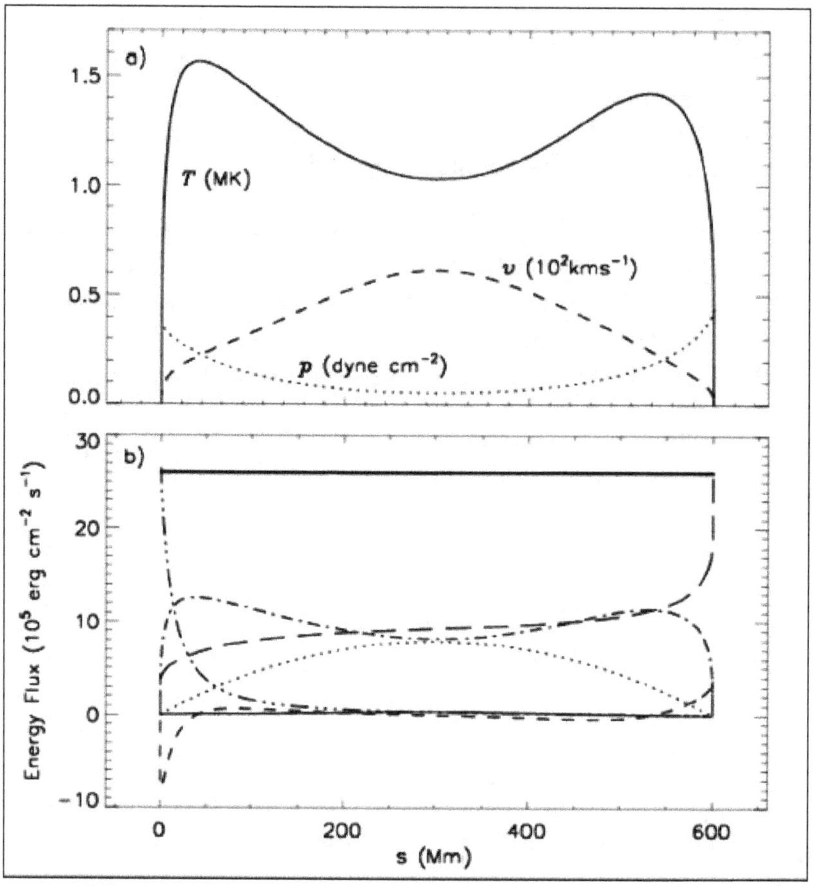

A modelled example of a quiescent coronal loop (energy contributions).

The idea that the coronal heating problem is solely down to some coronal heating mechanism is misleading. Firstly, the plasma filling over-dense loops is drained directly from the chromosphere. There is no coronal mechanism known that can compress coronal plasma and feed it into coronal loops at coronal altitudes. Secondly, observations of coronal upflows point to a chromospheric source of plasma. The plasma is therefore chromospheric in origin; there must be consideration of this when looking into coronal heating mechanisms. This is a *chromospheric energization* and *coronal heating phenomenon* possibly linked through a common mechanism.

Dynamic Flows

All of the above space missions have been highly successful in observing strong plasma flows and highly dynamic processes in coronal loops. For example, SUMER observations suggest flow velocities of 5–16 km/s in the solar disk, and other joint SUMER/TRACE observations detect flows of 15–40 km/s. Very high velocities have been detected by the Flat Crystal Spectrometer (FCS) on board the Solar Maximum Mission, where plasma velocities were found in the range of 40–60 km/s.

The new solar observatory Hinode (Solar-B), launched September 2006, will observe the magnetic structure of the corona.

SOLAR PHYSICS

Solar physics is the branch of astrophysics that specializes in the study of the Sun. It deals with detailed measurements that are possible only for our closest star. It intersects with many disciplines of pure physics, astrophysics, and computer science, including fluid dynamics, plasma physics including magnetohydrodynamics, seismology, particle physics, atomic physics, nuclear physics, stellar evolution, space physics, spectroscopy, radiative transfer, applied optics, signal processing, computer vision, computational physics, stellar physics and solar astronomy.

Because the Sun is uniquely situated for close-range observing (other stars cannot be resolved with anything like the spatial or temporal resolution that the Sun can), there is a split between the related discipline of observational astrophysics (of distant stars) and observational solar physics.

The study of solar physics is also important as it provides a "physical laboratory" for the study of plasma physics.

Babylonians were keeping a record of solar eclipses, with the oldest record originating from the ancient city of Ugarit, in modern-day Syria. This record dates to about 1300 BC. Ancient Chinese astronomers were also observing solar phenomena (such as solar eclipses and visible sunspots) with the purpose of keeping track of calendars, which were based on lunar and solar cycles. Unfortunately, records kept before 720 BC are very vague and offer no useful information. However, after 720 BC, 37 solar eclipses were noted over the course of 240 years.

Astronomical knowledge flourished in the Islamic world during medieval times. Many observatories were built in cities from Damascus to Baghdad, where detailed astronomical observations were taken. Particularly, a few solar parameters were measured and detailed observations of the Sun were taken. Solar observations were taken with the purpose of navigation, but mostly for timekeeping. Islam requires its followers to pray five times a day, at specific position of the Sun in the sky. As such, accurate observations of the Sun and its trajectory on the sky were needed. In the late 10th century, Iranian astronomer Abu-Mahmud Khojandi built a massive observatory near Tehran. There, he took accurate measurements of a series of meridian transits of the Sun, which he later used to calculate the obliquity of the ecliptic. Following the fall of the Western Roman Empire, Western Europe was cut from all sources of ancient scientific knowledge, especially those written in Greek. This, plus de-urbanisation and diseases such as the Black Death led to a decline in scientific knowledge in Medieval Europe, especially in the early Middle Ages. During this period, observations of the Sun were taken either in relation to the zodiac, or to assist in building places of worship such as churches and cathedrals.

In astronomy, the renaissance period started with the work of Nicolaus Copernicus. He proposed that planets revolve around the Sun and not around the Earth, as it was believed at the time. This model is known as the heliocentric model. His work was later expanded by Johannes Kepler and Galileo Galilei. Particularly, Galilei used his new telescope to look at the Sun. In 1610, he discovered sunspots on its surface. Modern day solar physics is focused towards understanding the many phenomena observed with the help of modern telescopes and satellites. Of particular interest are the structure of the solar photosphere, the coronal heat problem and sunspots.

SOLAR PHENOMENA

Solar phenomena are the natural phenomena occurring within the magnetically heated outer atmospheres in the Sun. These phenomena take many forms, including solar wind, radio wave flux, energy bursts such as solar flares, coronal mass ejection or solar eruptions, coronal heating and sunspots.

These phenomena are apparently generated by a helical dynamo near the center of the Sun's mass that generates strong magnetic fields and a chaotic dynamo near the surface that generates smaller magnetic field fluctuations.

The total sum of all solar fluctuations is referred to as solar variation. The collective effect of all solar variations within the Sun's gravitational field is referred to as space weather. A major weather component is the solar wind, a stream of plasma released from the Sun's upper atmosphere. It is responsible for the aurora, natural light displays in the sky in the Arctic and Antarctic. Space weather disturbances can cause solar storms on Earth, disrupting communications, as well as geomagnetic storms in Earth's magnetosphere and sudden ionospheric disturbances in the ionosphere. Variations in solar intensity also affect Earth's climate. These variations can explain events such as ice ages and the Great Oxygenation Event, while the Sun's future expansion into a red giant will likely end life on Earth.

The first extant description of the solar corona was in 968, while the earliest sunspot drawing was in 1128 and a solar prominence was described in 1185 in the Russian Chronicle of Novgorod. The invention of the telescope allowed major advances in understanding, allowing the first detailed observations in the 1600s. Solar spectroscopy began in the 1800s, from which properties of the solar atmosphere could be determined, while the creation of daguerreotypy led to the first solar photographs on 2 April 1845. Photography assisted in the study of solar prominences, granulation and spectroscopy. Early in the 20th century, interest in astrophysics surged in America. A number of new observatories were built with solar telescopes around the world. The 1931 invention of the coronagraph allowed the corona to be studied in full daylight.

Types

Coronal Mass Ejection

A coronal mass ejection (CME) is a massive burst of solar wind and magnetic fields rising above the solar corona. Near solar maxima, the Sun produces about three CMEs every day, whereas solar minima feature about one every five days. CMEs, along with solar flares of other origin, can disrupt radio transmissions and damage satellites and electrical transmission line facilities, resulting in potentially massive and long-lasting power outages.

Coronal mass ejections often appear with other forms of solar activity, most notably solar flares, but no causal relationship has been established. Most weak flares do not have CMEs; most powerful ones do. Most ejections originate from active regions on the Sun's surface, such as sunspot groupings associated with frequent flares. Other forms of solar activity frequently associcated with coronal mass ejections are eruptive prominences, coronal dimming, coronal waves and Moreton waves, also called solar tsunami.

Magnetic reconnection is responsible for CME and solar flares. Magnetic reconnection is the name given to the rearrangement of magnetic field lines when two oppositely directed magnetic fields are brought together. This rearrangement is accompanied with a sudden release of energy stored in the original oppositely directed fields.

When a CME impacts the Earth's magnetosphere, it temporarily deforms the Earth's magnetic field, changing the direction of compass needles and inducing large electrical ground currents in Earth itself; this is called a geomagnetic storm and it is a global phenomenon. CME impacts can induce magnetic reconnection in Earth's magnetotail (the midnight side of the magnetosphere); this launches protons and electrons downward toward Earth's atmosphere, where they form the aurora.

Flares

A solar flare is a sudden flash of brightness observed over the Sun's surface or the solar limb, which is interpreted as an energy release of up to 6×10^{25} joules of energy (about a sixth of the total Sun's energy output each second or 160 billion megatons of TNT equivalent, over 25,000 times more energy than released from the impact of Comet Shoemaker–Levy 9 with Jupiter). They may be followed by a coronal mass ejection. The flare ejects clouds of electrons, ions and atoms through the corona into space. These clouds typically reach Earth a day or two after the event. Similar phenomena in other stars are known as stellar flares.

Solar flares strongly influence space weather near the Earth. They can produce streams of highly energetic particles in the solar wind, known as a solar proton event. These particles can impact the Earth's magnetosphere in the form of a geomagnetic storm and present radiation hazards to spacecraft and astronauts.

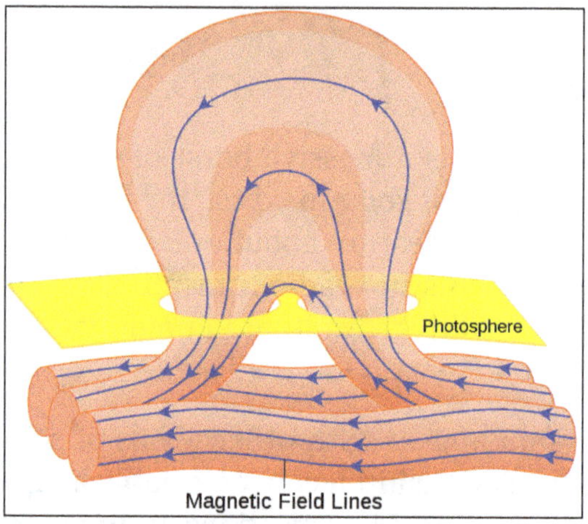

Diagram of the magnetic-field structure of a solar flare and its origin, inferred to result from the deformation of such a magnetic structure linking the solar interior with the solar atmosphere up through the corona.

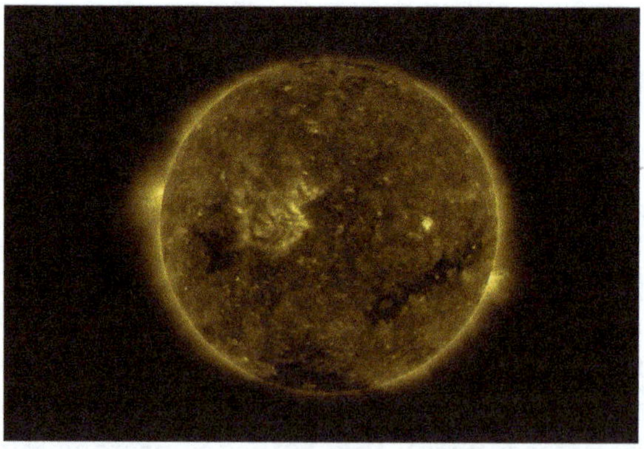

Irradiance

Irradiance is the power per unit area produced by the Sun in the form of electromagnetic radiation. Irradiance may be measured in space or at the Earth's surface after atmospheric absorption and scattering. Total solar irradiance (TSI), is a measure of the solar radiative power per unit area normal to the rays, incident on the Earth's upper atmosphere. The solar constant is a conventional measure of mean TSI at a distance of one Astronomical Unit (AU).

Insolation is a function of distance from the Sun, the solar cycle, and cross-cycle changes. Irradiance on Earth is most intense at points directly facing (normal to) the Sun.

Proton Event

Solar particles interact with Earth's magnetosphere.

A solar proton event (SPE), or "proton storm", occurs when particles (mostly protons) emitted by the Sun become accelerated either close to the Sun during a flare or in interplanetary space by CME shocks. The events can include other nuclei such as helium ions and HZE ions. These particles cause multiple effects. They can penetrate the Earth's magnetic field and cause ionization in the ionosphere. The effect is similar to auroral events, except that protons rather than electrons are involved. Energetic protons are a significant radiation hazard to spacecraft and astronauts. Energetic protons can reach Earth within 30 minutes of a major flare's peak.

Prominences and Filaments

A prominence is a large, bright, gaseous feature extending outward from the Sun's surface, often in the shape of a loop. Prominences are anchored to the Sun's surface in the photosphere and extend outwards into the corona. While the corona consists of high temperature plasma, which do not emit much visible light, prominences contain much cooler plasma, similar in composition to that of the chromosphere.

Prominence plasma is typically a hundred times cooler and denser than coronal plasma. A prominence forms over timescales of about an earthly day and may persist for weeks or months. Some prominences break apart and from CMEs.

A typical prominence extends over many thousands of kilometers; the largest on record was estimated at over 800,000 kilometres (500,000 mi) long – roughly the solar radius.

When a prominence is viewed against the Sun instead of space, it appears darker than the background. This formation is called a solar filament. It is possible for a projection to be both a filament and a prominence. Some prominences are so powerful that they eject matter at speeds ranging from 600 km/s to more than 1000 km/s. Other prominences form huge loops or arching columns of glowing gases over sunspots that can reach heights of hundreds of thousands of kilometers. Prominences may last for days or months.

Wind

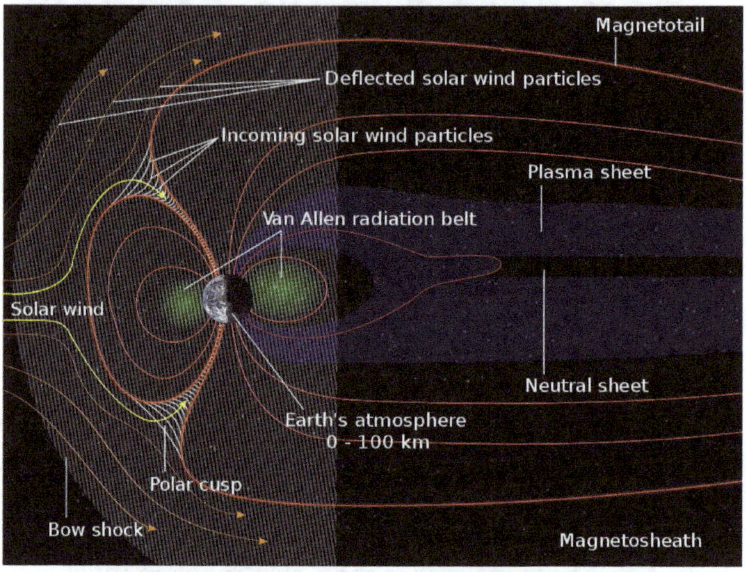

Schematic of Earth's magnetosphere. The solar wind flows from left to right.

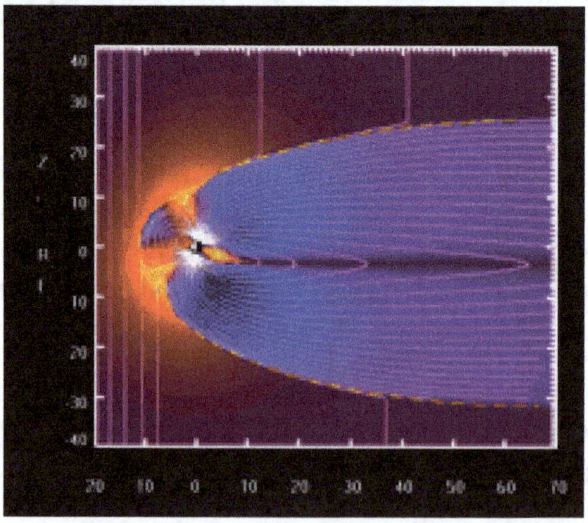

Simulation of Earth's magnetic field in interaction with (solar) interplanetar magnetic field that illustrates the dynamical changes of the global magnetic field in the course of a disturbance: a temporary compression of the magnetosphere by enhanced flow of the solar wind is followed by a tailward stretching of the field lines.

The solar wind is a stream of plasma released from the Sun's upper atmosphere. It consists of mostly electrons and protons with energies usually between 1.5 and 10 keV. The stream of particles varies in density, temperature and speed over time and over solar longitude. These particles can escape the Sun's gravity because of their high energy.

The solar wind is divided into the slow solar wind and the fast solar wind. The slow solar wind has a velocity of about 400 kilometres per second (250 mi/s), a temperature of $1.4–1.6\times10^6$ K and a composition that is a close match to the corona. The fast solar wind has a typical velocity of 750 km/s, a temperature of 8×10^5 K and nearly matches the photospher's. The slow solar wind is twice as dense and more variable in intensity than the fast solar wind. The slow wind has a more complex structure, with turbulent regions and large-scale organization.

Both the fast and slow solar wind can be interrupted by large, fast-moving bursts of plasma called interplanetary CMEs, or ICMEs. They cause shock waves in the thin plasma of the heliosphere, generating electromagnetic waves and accelerating particles (mostly protons and electrons) to form showers of ionizing radiation that precede the CME.

Sunspot

Sunspots are darker, cooler areas on the surface of the sun in a region called the photosphere. Sunspots form on the surface of the Sun due to strong magnetic field lines coming up from within the Sun trough the solar surface and appear visibly as dark spots compared to their surroundings. These sunspots which can become many times bigger than the Earth are always dark because they are much cooler than the surrounding surface of the Sun itself. A big sunspot can have a temperature of 3700°C. This sounds like much but if we compare this with the temperature of the photosphere of the Sun which is about 5500°C, then you see that there is a considerable difference. As a matter of fact, if we could take a sunspot out of the Sun and place it into our night sky it would only be as bright as the full moon, a very big contrast with the bright Sun itself.

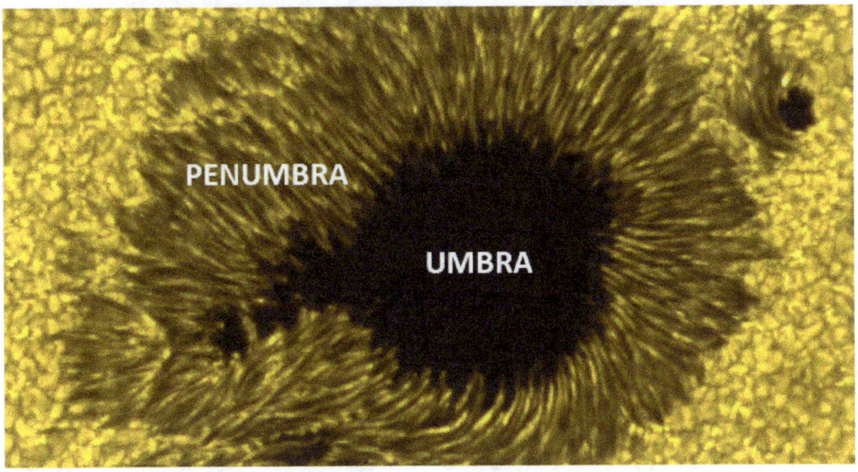

Sunspots are a common sight on our Sun during the years around solar maximum. Solar maximum or solar max is the period of greatest solar activity in the solar cycle of the Sun, where one solar cycle lasts about 11 years. Around solar minimum, only very few or even no sunspots can be found. Sunspots form where magnetic field lines come up from the Sun's interior trough the solar surface meaning that every sunspot has it's own polarity.

A sunspot consists of two parts:

- The dark part (umbra).
- Lighter part around the dark part (penumbra).

Sunspot Groups

The image below shows a large and complex group with numerous sunspots. This is called a sunspot group or an Active Region. Every day, all the sunspot groups on the Earth-facing solar disk are analysed for their eruptive threat and receive a number. This is done by the NOAA Space Weather Prediction Center. Groups of sunspots, especially those with complex magnetic layouts, are know to cause a phenomenon called solar flares.

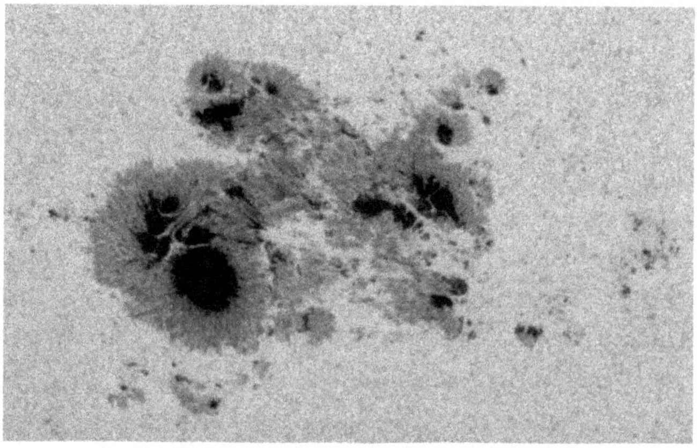

Rotation of the Sun

The Sun rotates around its axis just like Earth does. Solar features on the Sun like sunspot regions follow the rotation of the Sun. This means that a sunspot region travels across the solar disk from east to west as seen from Earth. This is important because sunspot regions need to be close to the central meridian (as seen from Earth) in order to be able to send coronal mass ejection towards Earth. It takes a sunspot region near the equator about 2 weeks to move from the east limb to west limb as seen from Earth. The further away a sunspot region is from the equator the longer it takes move across the face of the Sun. This is because the Sun rotates faster at its equator than at its poles. The rotational period is approximately 25.6 days at the equator and 33.5 days at the poles. Viewed from Earth as it orbits the Sun, the apparent rotational period of the Sun at its equator is about 28 days.

Sunspot Cycle

The number of sunspots observed on the "surface" of the Sun varies from year to year. This rise and fall in sunspot counts varies in a cyclical way; the length of the cycle is around eleven years on average. The cyclical variation in sunspot counts, discovered in 1843 by the amateur German astronomer Samuel Heinrich Schwabe, is called "the Sunspot Cycle".

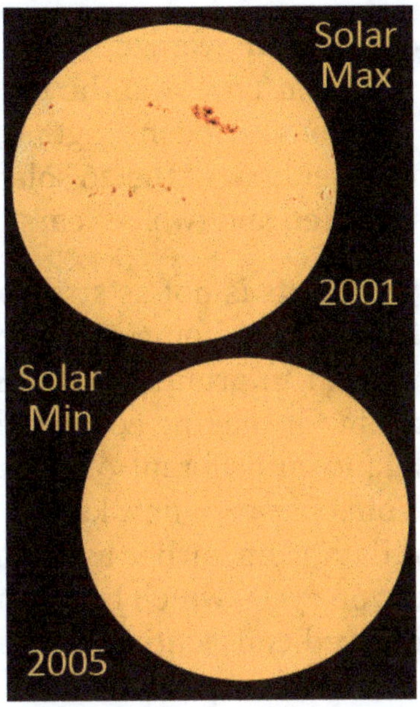

These two images of the Sun show how the number of sunspots varies over the course of a sunspot cycle. The above, with many sunspots, was taken near solar max in March 2001. The below image, in which no spots are evident, was taken near solar min in January 2005.

Solar Max and Solar Min

A peak in the sunspot count is referred to as a time of "solar maximum" (or "solar max"), whereas a period when few sunspots appear is called a "solar minimum" (or "solar min"). An example of a recent sunspot cycle spans the years from the solar min in 1986, when 13 sunspots were seen, through the solar max in 1989 when more than 157 sunspots appeared, on to the next solar min in 1996 (ten years after the 1986 solar min) when the sunspot count had fallen back down to fewer than 9.

Along with the number of sunspots, the location of sunspots varies throughout the sunspot cycle. At solar min, sunspots tend to form around latitudes of 30° to 45° North and South of the Sun's equator. As the solar cycle progresses through solar max, sunspots tend to appear closer to the equator, around a latitude of 15°. Towards the end of a cycle, with solar min once again approaching, sunspots form quite close to the solar equator, around 7° North and South latitude. There is often an overlap in this latitudinal migration trend around solar min, when sunspots of the outgoing cycle are forming at low latitudes and sunspots of the upcoming cycle begin to form at high latitudes once again. This gradual equatorward drift of sunspots throughout the sunspot cycle, which was first noticed in the early 1860's by the German astronomer Gustav Spörer and the Englishman Richard Christopher Carrington, is often called Spörer's Law. In 1904 another English astronomer, Edward Walter Maunder, constructed the first "butterfly diagram", a graphical plot of this sunspot migration trend.

11-Year Cycle

The duration of the sunspot cycle is, on average, around eleven years. However, the length of the cycle does vary. Between 1700 and the present, the sunspot cycle (from one solar min to the next solar min) has varied in length from as short as nine years to as long as fourteen years. Note, however, that of the 26 solar cycles during that three-century span, 21 had a length between ten and twelve years.

Arriving at a precise count of sunspots is not as straightforward as it might appear. Some spots are much larger than others, some sunspots partially merge together at their edges, and many spots appear in groups. In 1848 a Swiss astronomer named Rudolf Wolf devised an algorithm for making consistent counts of sunspots that allows comparisons between data from different observers across the centuries. The sunspot count derived using Wolf's formula, now known as the Wolf sunspot number, is still in use today. Wolf used data from earlier astronomers to reconstruct sunspot counts as far back as the 1755-1766 cycle, which he dubbed "cycle 1". Since then, subsequent cycles have been numbered consecutively, so the cycle that began with the 2008 solar minimum is cycle 24.

Sunspots Mark Magnetic Disturbances

The Sun is typically very active when sunspot counts are high. Sunspots are indicators of disturbances in the Sun's magnetic field, which can generate energetic solar events like solar flares and coronal mass ejections. Since reasonably reliable records of sunspot counts extend back to the early 1700s, long before other measures of solar activity could be observed, sunspot counts serve as a valuable, relatively long-term indicator of solar activity. The Sun emits significantly more radiation than usual in the X-ray and ultraviolet portions of the electromagnetic spectrum during solar max, and this extra energy significantly alters the uppermost layers of Earth's atmosphere.

The 11-year sunspot cycle is actually half of a longer, 22-year cycle of solar activity. Each time the sunspot count rises and falls, the magnetic field of the Sun associated with sunspots reverses polarity; the orientation of magnetic fields in the Sun's northern and southern hemispheres switch. Thus, in terms of magnetic fields, the solar cycle is only complete (with the fields back the way they were at the start of the cycle) after two

11-year sunspot cycles. This solar cycle is, on average, about 22 years long - twice the duration of the sunspot cycle.

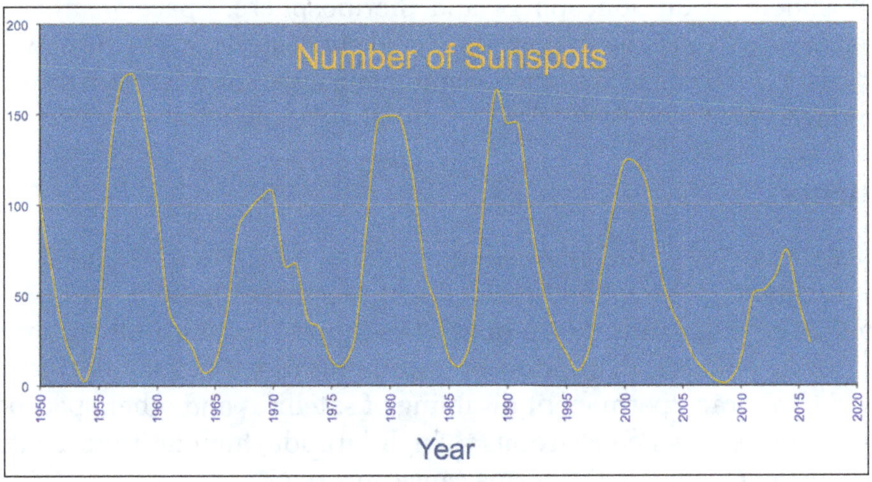

This graph shows the number of sunspots counted each year for several decades. Notice how the sunspot count rises and falls in an 11-year cycle.

Some scientists believe there is evidence for other, longer-period variations in the sunspot and solar cycles. Other scientists are skeptical about such claims. Most scientists think we need more data, spanning longer periods of time, to definitively resolve this issue. Besides these regular cycles, the Sun has exhibited periods of very unusual sunspot counts. Most notably, from about 1645 to 1715 there were very few sunspots - in some years none at all were observed. This period, now called the Maunder Minimum (after E.W. Maunder, who did important pioneering work related to this phenomenon), corresponded to an extremely cold spell in Europe known as the Little Ice Age.

Effects

Space Weather

Aurora australis in the Earth's atmosphere observed by Space Shuttle Discovery.

Space weather is the environmental condition within the Solar System, including the solar wind. It is studied especially surrounding the Earth, including conditions from the magnetosphere to the ionosphere and thermosphere. Space weather is distinct from terrestrial weather of the (troposphere and stratosphere). The term was not used until the 1990s. Prior to that time, such phenomena were considered to be part of physics or aeronomy.

Solar Storms

Solar storms are caused by disturbances on the Sun, most often coronal clouds associated with solar flare CMEs emanating from active sunspot regions, or less often from coronal holes. The Sun can produce intense geomagnetic and proton storms capable of causing power outages, disruption or communications blackouts (including GPS systems) and temporary/permanent disabling of satellites and other spaceborne technology. Solar storms may be hazardous to high-latitude, high-altitude aviation and to human spaceflight. Geomagnetic storms cause aurorae.

The most significant known solar storm occurred in September 1859 and is known as the Carrington event.

Aurora

An aurora is a natural light display in the sky, especially in the high latitude (Arctic and Antarctic) regions, in the form of that forms a large circle around the pole. They are caused by the collision of solar wind and charged magnetospheric particles with the high altitude atmosphere (thermosphere).

Most auroras occur in a band known as the auroral zone, which is typically 3° to 6° wide in latitude and observed at 10° to 20° from the geomagnetic poles at all longitudes, but often most vividly around the spring and autumn equinoxes. The charged particles and solar wind are directed into the atmosphere by the Earth's magnetosphere. A geomagnetic storm expands the auroral zone to lower latitudes.

Auroras are associated with the solar wind. The Earth's magnetic field traps its particles, many of which travel toward the poles where they are accelerated toward Earth. Collisions between these ions and the atmosphere release energy in the form of auroras appearing in large circles around the poles. Auroras are more frequent and brighter during the solar cycle's intense phase when CMEs increase the intensity of the solar wind.

Geomagnetic Storm

A geomagnetic storm is a temporary disturbance of the Earth's Magnetosphere caused by a solar wind shock wave and/or cloud of magnetic field that interacts with the Earth's magnetic field. The increase in solar wind pressure compresses the magnetosphere and the solar wind's magnetic field interacts with the Earth's magnetic field to transfer increased energy into the magnetosphere. Both interactions increase plasma movement through the magnetosphere (driven by increased electric fields) and an increase in electric current in the magnetosphere and ionosphere.

The disturbance in the interplanetary medium that drives a storm may be due to a CME or a high speed stream (co-rotating interaction region or CIR) of the solar wind originating from a region of weak magnetic field on the solar surface. The frequency of geomagnetic storms increases and decreases with the sunspot cycle. CME driven storms are more common during the solar maximum of the solar cycle, while CIR-driven storms are more common during the solar minimum.

Several space weather phenomena are associated with geomagnetic storms. These include Solar Energetic Particle (SEP) events, geomagnetically induced currents (GIC), ionospheric disturbances that cause radio and radar scintillation, disruption of compass navigation and auroral displays at much lower latitudes than normal. A 1989 geomagnetic storm energized ground induced currents that disrupted electric power distribution throughout most of the province of Quebec and caused aurorae as far south as Texas.

Sudden Ionospheric Disturbance

A sudden ionospheric disturbance (SID) is an abnormally high ionization/plasma density in the D region of the ionosphere caused by a solar flare. The SID results in a sudden increase in radio-wave absorption that is most severe in the upper medium frequency (MF) and lower high frequency (HF) ranges, and as a result often interrupts or interferes with telecommunications systems.

Geomagnetically Induced Currents

Geomagnetically induced currents are a manifestation at ground level of space weather, which affect the normal operation of long electrical conductor systems. During space weather events, electric currents in the magnetosphere and ionosphere experience large variations, which manifest also in the Earth's magnetic field. These variations induce currents (GIC) in earthly conductors. Electric transmission grids and buried pipelines are common examples of such conductor systems. GIC can cause problems such as increased corrosion of pipeline steel and damaged high-voltage power transformers.

Carbon-14

The production of carbon-14 (radiocarbon: ^{14}C) is related to solar activity. Carbon-14 is produced in the upper atmosphere when cosmic ray bombardment of atmospheric nitrogen (^{14}N) induces the Nitrogen to undergo $\beta+$ decay, thus transforming into an unusual isotope of carbon with an atomic weight of 14 rather than the more common 12. Because galactic cosmic rays are partially excluded from the Solar System by the outward sweep of magnetic fields in the solar wind, increased solar activity reduces ^{14}C production.

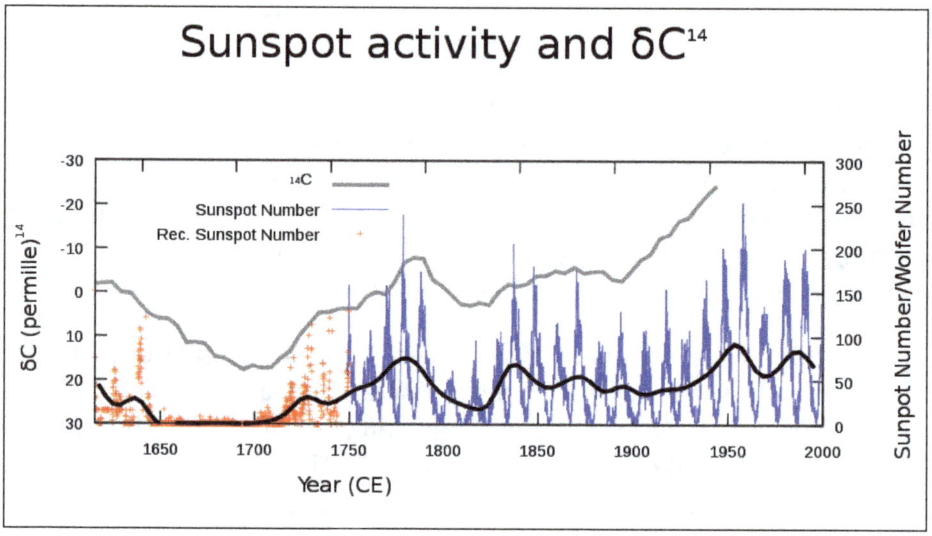

Sunspot record (blue) with ^{14}C (inverted).

Atmospheric ^{14}C concentration is lower during solar maxima and higher during solar minima. By measuring the captured ^{14}C in wood and counting tree rings, production of radiocarbon relative to recent wood can be measured and dated. A reconstruction of the past 10,000 years shows that the ^{14}C production was much higher during the mid-Holocene 7,000 years ago and decreased until 1,000 years ago. In addition to variations in solar activity, long term trends in carbon-14 production are influenced by changes in the Earth's geomagnetic field and by changes in carbon cycling within the biosphere (particularly those associated with changes in the extent of vegetation between ice ages).

Solar Cycle

The solar cycle or solar magnetic activity cycle is a nearly periodic 11-year change in the Sun's activity. Levels of solar radiation and ejection of solar material, the number and size of sunspots, solar flares, and coronal loops all exhibit a synchronized fluctuation, from active to quiet to active again, with a period of 11 years. This cycle has been observed for centuries by changes in the Sun's appearance and by terrestrial phenomena such as auroras.

The changes on the Sun cause effects in space, in the Earth's atmosphere, and on Earth's surface. While the cycle is the dominant influence on solar activity, aperiodic fluctuations also occur.

Solar cycles have an average duration of about 11 years. Solar maximum and solar minimum refer to periods of maximum and minimum sunspot counts. Cycles span from one minimum to the next.

Patterns

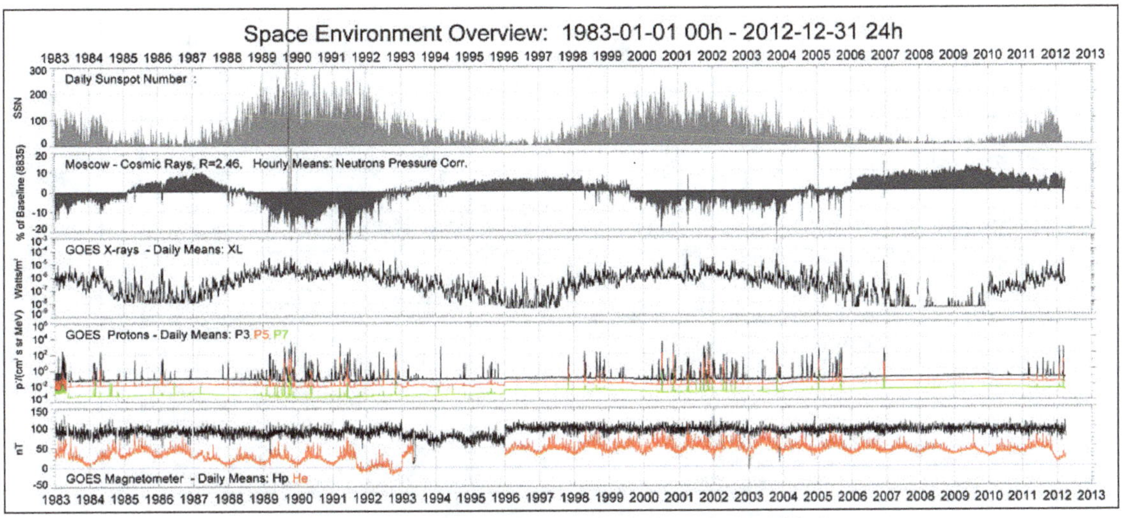

An overview of three solar cycles shows the relationship between the sunspot cycle, galactic cosmic rays, and the state of our near-space environment.

The Waldmeier effect names the observation that cycles with larger maximum amplitudes tend to take less time to reach their maxima than cycles with smaller amplitudes; maximum amplitudes are negatively correlated to the lengths of earlier cycles, aiding prediction.

Solar maxima and minima also exhibit fluctuations at time scales greater than solar cycles. Increasing and decreasing trends can continue for periods of a century or more.

The Schwabe Cycle is thought to be an amplitude modulation of the 87 year (70–100 year) Gleissberg cycle, named after Wolfgang Gleißberg. The Gleisberg cycle implied that the next solar cycle have a maximum smoothed sunspot number of about 145±30 in 2010 (instead 2010 was just after the cycle's solar minimum) and that the following cycle have a maximum of about 70±30 in 2023.

Associated centennial variations in magnetic fields in the Corona and Heliosphere have been detected using Carbon-14 and beryllium-10 cosmogenic isotopes stored in terrestrial reservoirs such as ice sheets and tree rings and by using historic observations of Geomagnetic storm activity, which bridge the time gap between the end of the usable cosmogenic isotope data and the start of modern satellite data.

These variations have been successfully reproduced using models that employ magnetic flux continuity equations and observed sunspot numbers to quantify the emergence of magnetic flux from the top of the solar atmosphere and into the Heliosphere, showing that sunspot observations, geomagnetic activity and cosmogenic isotopes offer a convergent understanding of solar activity variations.

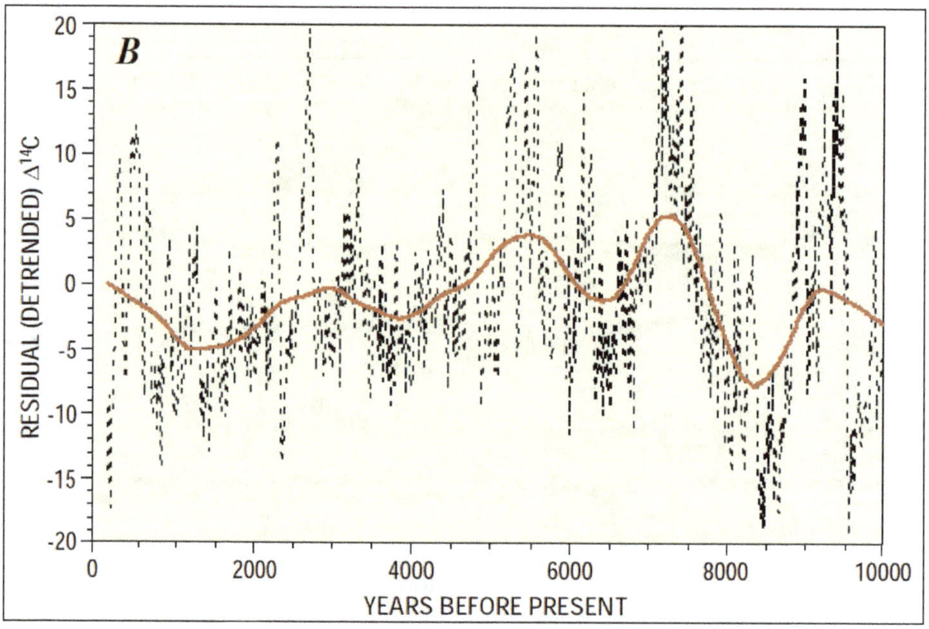

2,300 year Hallstatt solar variation cycles.

Hypothesized Cycles

Periodicity of solar activity with periods longer than the sunspot cycle has been proposed, including: The 210 year Suess cycle (a.k.a. "de Vries cycle", named after Hans Eduard Suess and Hessel De Vries respectively) is recorded from radiocarbon studies, although "little evidence of the Suess Cycle" appears in the 400-year sunspot record.

The Hallstatt cycle (named after a cool and wet period in Europe when glaciers advanced) is hypothesized to extend for approximately 2,400 years. And as yet unnamed cycle may extend over 6,000 years.

In carbon-14 cycles of 105, 131, 232, 385, 504, 805 and 2,241 years have been observed, possibly matching cycles derived from other sources. Damon and Sonett proposed carbon 14-based medium- and short-term variations of periods 208 and 88 years; as well as suggesting a 2300-year radiocarbon period that modulates the 208-year period.

During the Upper Permian 240 million years ago, mineral layers created in the Castile Formation show cycles of 2,500 years.

Solar Magnetic Field

The Sun's magnetic field structures its atmosphere and outer layers all the way through the corona and into the solar wind. Its spatiotemporal variations lead to various measurable solar phenomena. Other solar phenomena are closely related to the cycle, which serves as the energy source and dynamical engine for the former.

Effects

Solar

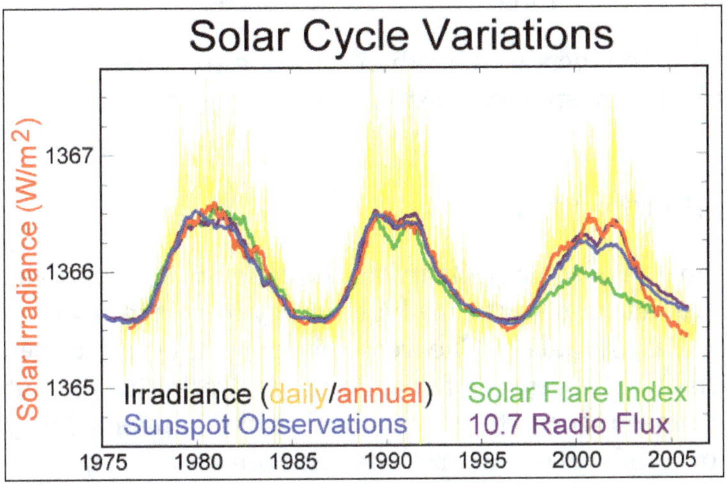

Activity cycles 21, 22 and 23 seen in sunspot number index, TSI, 10.7cm radio flux, and flare index. The vertical scales for each quantity have been adjusted to permit overplotting on the same vertical axis as TSI. Temporal variations of all quantities are tightly locked in phase, but the degree of correlation in amplitudes is variable to some degree.

Surface Magnetism

Sunspots eventually decay, releasing magnetic flux in the photosphere. This flux is dispersed and churned by turbulent convection and solar large-scale flows. These transport mechanisms lead to the accumulation of magnetized decay products at high solar latitudes, eventually reversing the polarity of the polar fields (notice how the blue and yellow fields reverse in the Hathaway/NASA/MSFC graph above).

The dipolar component of the solar magnetic field reverses polarity around the time of solar maximum and reaches peak strength at the solar minimum.

Space

Spacecraft

CMEs (coronal mass ejections) produce a radiation flux of high-energy protons, sometimes known as solar cosmic rays. These can cause radiation damage to electronics and solar cells in satellites. Solar proton events also can cause single-event upset (SEU) events on electronics; at the same, the reduced flux of galactic cosmic radiation during solar maximum decreases the high-energy component of particle flux.

CME radiation is dangerous to astronauts on a space mission who are outside the shielding produced by the Earth's magnetic field. Future mission designs (e.g., for a

Mars Mission) therefore incorporate a radiation-shielded "storm shelter" for astronauts to retreat to during such an event.

Gleißberg developed a CME forecasting method that relies on consecutive cycles.

On the positive side, the increased irradiance during solar maximum expands the envelope of the Earth's atmosphere, causing low-orbiting space debris to re-enter more quickly.

Galactic Cosmic Ray Flux

The outward expansion of solar ejecta into interplanetary space provides overdensities of plasma that are efficient at scattering high-energy cosmic rays entering the solar system from elsewhere in the galaxy. The frequency of solar eruptive events is modulated by the cycle, changing the degree of cosmic ray scattering in the outer solar system accordingly. As a consequence, the cosmic ray flux in the inner solar system is anticorrelated with the overall level of solar activity. This anticorrelation is clearly detected in cosmic ray flux measurements at the Earth's surface.

Some high-energy cosmic rays entering Earth's atmosphere collide hard enough with molecular atmospheric constituents that they occasionally cause nuclear spallation reactions. Fission products include radionuclides such as 14C and 10Be that settle on the Earth's surface. Their concentration can be measured in ice cores, allowing a reconstruction of solar activity levels into the distant past. Such reconstructions indicate that the overall level of solar activity since the middle of the twentieth century stands amongst the highest of the past 10,000 years, and that epochs of suppressed activity, of varying durations have occurred repeatedly over that time span.

Solar Dynamo

The 11-year sunspot cycle is half of a 22-year Babcock–Leighton solar dynamo cycle, which corresponds to an oscillatory exchange of energy between toroidal and poloidal solar magnetic fields. At solar-cycle maximum, the external poloidal dipolar magnetic field is near its dynamo-cycle minimum strength, but an internal toroidal quadrupolar field, generated through differential rotation within the tachocline, is near its maximum strength. At this point in the dynamo cycle, buoyant upwelling within the Convection zone forces emergence of the toroidal magnetic field through the photosphere, giving rise to pairs of sunspots, roughly aligned east–west with opposite magnetic polarities. The magnetic polarity of sunspot pairs alternates every solar cycle, a phenomenon known as the Hale cycle.

During the solar cycle's declining phase, energy shifts from the internal toroidal magnetic field to the external poloidal field, and sunspots diminish in number. At solar minimum, the toroidal field is, correspondingly, at minimum strength, sunspots are relatively rare and the poloidal field is at maximum strength. During the next cycle, differential rotation converts magnetic energy back from the poloidal to the toroidal field, with a polarity that is opposite to the previous cycle. The process carries on continuously, and in an idealized, simplified scenario, each 11-year sunspot cycle corresponds to a change in the polarity of the Sun's large-scale magnetic field.

Although the tachocline has long been thought to be the key to generating the Sun's large-scale magnetic field, recent research has questioned this assumption. Radio observations of brown dwarfs have indicated that they also maintain large-scale magnetic fields and may display cycles of magnetic activity. The Sun has a radiative core surrounded by a convective envelope, and at the boundary of these two is the tachocline. However, brown dwarfs lack radiative cores and tachoclines. Their structure consists of a solar-like convective envelope that exists from core to surface. Since they lack a tachocline yet still display solar-like magnetic activity, it has been suggested that solar magnetic activity is only generated in the convective envelope.

Speculated Influence of the Planets

It has long been theorized that the planets may have an influence on the solar cycle, with many speculative papers published through years. For example, In 2012 a team led by José Abreu calculated the torque exerted by the planets on a non-spherical tachocline layer deep in the Sun and proposed an explanation of how the tiny tidal force can synchronize the solar dynamo. However, their results were shown to be an artifact of the incorrectly applied smoothing method leading to the aliasing. Moreover, the solar variability is known to be essentially stochastic and unpredictable beyond one solar cycle, which contradicts the idea of the deterministic planetary influence on solar dynamo.

STANDARD SOLAR MODEL

The standard solar model is one of the most complete and successful theories in modern astronomy.

The Sun is the most prominent object in our sky, and understanding it has been one of the primary focuses since the birth of modern astronomy. Additionally, after the realization that the Sun was itself a star, solar research became a means for investigating conditions too distant to observe directly. Starting in the early twentieth century, astronomers began building a model of the Sun that would accurately predict the Sun's observable features. If the model can accurately predict what is observed, then it is reasonable to assume that it can accurately tell astronomers about what they cannot observe, both inside the Sun and its behavior at other epochs. This model is known as the standard solar model, and it has been in a state of constant evolution since its inception.

The standard solar model has four basic assumptions, the first being that the sun evolves in hydrostatic equilibrium. Hydrostatic equilibrium implies a local balance between pressure and gravity, can be expressed as:

$$\frac{dP}{dr} = -\frac{Gm\rho}{r^2}$$

where P is the pressure, ρ is the density, and m is the mass contained inside the radius r. To describe this condition in detail, one must specify the temperature, density, and composition of the stellar material. This combination is known as an equation of state, the simplest of which being the ideal gas law:

$$p = \frac{\rho TR}{\mu}$$

where μ is the mean molecular weight, T is the temperature of the material, and is a constant.

The second assumption of the standard solar model is that energy can be transferred in the star via radiation, conduction, convection, and neutrino losses. Radiation and convection are the dominant forms of energy transport, and a measure of energy flow is the temperature gradient produced through each process:

$$\frac{dT}{dr} = -\frac{3\kappa\rho L}{16\pi acr^2T^3}$$

and

$$\frac{dT}{dr} = \left(1 - \frac{1}{\gamma}\right)\left(\frac{T}{P}\right)\frac{dp}{dr}$$

for radiative and conductive transport, respectively, where κ is the opacity of the stellar material, L is the luminosity, and γ is the ratio of specific heats, C_p/C_V. These two processes vary in their efficiencies depending on the local values at a given point inside the star.

The third assumption of the model is that thermonuclear reactions are the only source of energy production inside the star. The thermonuclear reactions that take place in the core of a star like our Sun are processes that fuse hydrogen nuclei into helium nuclei, releasing copious quantities of energy in the process. There are two processes, or chains of reactions, that are responsible for the fusion in our Sun, these are the proton-proton (*pp*) chain and the carbon-nitrogen-oxygen (*CNO*) chain. These chains are described in more detail below. Fusion reactions require high densities and temperatures to take place, and therefore are predominantly found d in stellar cores. A measure of the luminosity in star created by these reactions is given by:

$$\frac{dL}{dr} = 4\pi r^2 \rho\varepsilon,$$

where ε is the energy production rate [ergs g^{-1} s^{-1}].

The final assumption of the standard solar model is that the sun was initially of a homogeneous, primordial composition, and highly convective at its main sequence turn on. Since heavy elements are neither created nor destroyed in the thermonuclear reactions in a solar-type star, they provide a record of the initial abundances, and only the relative amounts of hydrogen and 4helium are an indicator of stellar evolution. The standard solar model has been successful over its lifetime in reproducing the conditions observed today and producing scenarios for stellar interiors that are in agreement with standard physics and subsequent measurements. For example, Turck-Chieze define their model successful if it converges to solar values with errors less than given in the table below.

Quantity	Acceptable Error
Solar Luminosity, $L_\odot$	$\pm 5 \times 10^{-3} L_\odot$
Solar Age, $t_\odot$	$\pm 0.1 \times 10^{9}\ yr$
Solar Radius, $R_\odot$	$\pm 5 \times 10^{-4} R_\odot$

Until very recently, the biggest shortcoming of the standard solar model was the solar neutrino prediction. Experiments on the earth set an upper bound for the number of neutrinos that could be produced in the sun, and these numbers fell well short of the number of neutrinos predicted by the standard model. Solving this problem was the focus of solar model research over the last 30 years. The answer was recently found by $p = \frac{\rho TR}{\mu}$ when they measured oscillations in the type of neutrinos emitted from the sun.

Standard Solar Model Data

The next step is to insert the model assumptions into a code that will generate values for mass, luminosity, pressure, density, and composition for each value of stellar radius. These essential physical quantities can be combined in various ways to shed light on the processes occurring to create the physical conditions inside the star.

Composition

Equation $p = \frac{\rho TR}{\mu}$ can be inverted to solve for the mean molecular weight for a given region, which allows one to observe the overall change in composition throughout the star. This figure shows the decline in molecular weight with increasing radius. This decline is a product of the thermonuclear reactions that are powering the star. As mentioned above, during stellar evolution, hydrogen is converted into helium in the hot, dense core. That the star is no longer of a homogeneous composition is a clear indication of its evolution. The 4 helium content in the core of the star is enhanced as the star uses up its fuel, which can be seen from the nuclear chains mentioned above. The dominant process for converting hydrogen to helium in a solar-type star is the pp chain, where:

$$p + p \rightarrow D + e^{+} + \upsilon_{e}$$
$$D + p \rightarrow^{3} He + \gamma$$
$$^{3}He +^{3} He \rightarrow^{4} He + p + p$$

is known as pp-I, and is responsible for 84.6% of solar energy generation, where p are protons, D are deuterium nuclei, and ν and γ are emitted neutrinos and photons, respectively. The star has another option, pp-II, which occurs if:

$$^{3}He +^{4} He \rightarrow^{4} Be + \gamma$$

followed by further steps leading to the creation of ^{4}helium. This process produces 13.8% of solar energy.

The other process mentioned above is the CNO chain, which also converts hydrogen into ^{4}helium, but uses carbon, nitrogen, and oxygen nuclei as catalysts. This process does not dominate until 20-25 million degrees Kelvin, so in our Sun, where the core temperature is > 16 million Kelvin, this process is only secondary, producing the remaining 1.6% of solar energy. The CNO chain plays another important role in the core of a star, which is to change its composition. Carbon is converted into nitrogen through the CNO process, which should be observed as an enhancement of nitrogen and a depletion of carbon in the core. Both carbon and nitrogen go through the catalytic cycle about a dozen times over the life of the Sun, whereas oxygen nuclei only go through a fraction of a cycle.

Energy Production Regions

As discussed above, the thermonuclear reactions that power the sun require the high temperatures and densities associated with the core of the Sun. Equation $\frac{dL}{dr} = 4\pi r^{2} \rho \varepsilon$ can be rewritten to solve for the energy production rate. The energy production rate is seen to fall off appreciably by 0.2 $R_{\odot}$. This fact is further seen by noticing that the Sun's luminosity has reached its full value by 0.2 $R_{\odot}$, telling us that all of the energy production took place inside that radius. Similarly, Bahcall et al find 95% of solar luminosity produced within 0.21 $R_{\odot}$, and a similar calculation of energy production finds its peak at 0.09 Rwith half-maximum at 0.16 $R_{\odot}$. Another interesting question is to ask how different is the Sun today than at its main sequence turn on. This can be answered by taking a successful model and comparing its initial conditions with present values. Turck-Chieze et al find that since main sequence turn on, the Sun's luminosity has increased by 30%, the radius has increased by 15%, the core temperature has increased by 8%, the core pressure has increased by 61%, and the central hydrogen abundance has been depleted by 50%.

Convection

In the outer regions of a star, local perturbations may cause discrete regions to become unstable. The region becomes unstable when its density becomes appreciably different than its surroundings. This density change will result in a buoyancy force acting on the region and causing it to rise or fall. This bulk motion of material is convection, and is an important process that influences stellar structure by carrying energy and mixing stellar material. If we assume that the region changes adiabatically, and that the surrounding region is predominantly transporting energy via radiation, we can write down a condition for the region to be stable:

$$\nabla_{rad} < \nabla_{adb}$$

which is known as the Schwarzchild stability criterion. The ∇s are temperature gradients, and are defined as:

$$\nabla_{rad} = \frac{3\kappa LP}{16\pi acGmT^4}$$

the radiative temperature gradient, and

$$\nabla_{adb} = \nabla_{region}\left(\frac{d\,In\,T}{d\,In\,P}\right)_{region}$$

the adiabatic temperature gradient. When the stability criterion is violated (ie-when the radiative temperature gradient is greater than the adiabatic temperature gradient), the region will become unstable and convection will begin. When convection is most efficient, the adiabatic temperature gradient is roughly equal to the total gradient $\left(ie - \nabla_{adb} \approx \nabla_{tot}\right)$. These gradients can be calculated and one can determine where convection will play a significant role in the stellar structure. Convection is seen to be transporting its maximum fraction of stellar energy near $0.7\,R_{\odot}$.

SOLAR IRRADIANCE

Solar irradiation is the amount of electromagnetic radiation received from the sun per unit area (usually square meters). In other words, it's the amount of the sun's power detected by a measuring instrument. When these data are integrated over time, the information is called solar irradiation, insolation, or solar exposure. The amount of solar irradiance varies depending on how far the object is from the sun, the angle of the sun, and the solar cycle—the change in the sun's appearance and activity every 11 years. Irradiance can be measured for the moon, stars, or any other glowing object.

Breaking solar irradiance data up into its individual wavelengths (colors) of light, also called spectral irradiance, gives scientists additional information. The irradiance of the sun is studied at several wavelengths including visible light, infrared (IR), ultraviolet (UV), extreme ultraviolet (EVU), and X-rays.

Studying spectral irradiance is important because each wavelength is absorbed in various parts of the atmosphere. For example, the radiation from visible and IR light warms surfaces like the skin and the roofs of buildings. Also, changes in solar EVU output effects space weather which is an important concern for spacecraft and space travel.

The irradiance at the top of the Earth's atmosphere is about 1361 W/m^2. After passing through the atmosphere and losing energy, irradiation at the surface of the planet is 1000 W/m^2 on a clear day at sea level. The average daily irradiance of the Earth is about 6 kWh/m^2. Even small changes in the sun's irradiance can have dramatic effects on the climate, atmosphere, and ionosphere.

Applications of Solar Irradiance

Calculations using solar irradiation values are used to plan solar power systems. Most countries have insolation maps and tables for the last 30 to 60 years for use in planning. There are several solar power technologies the choose from and each uses a different portion of the total irradiation. Concentrated solar power uses only direct irradiation from the sun to generate electricity and is suitable for areas with little cloud cover. Solar voltaic panels are more versatile and use either direct or diffuse radiation.

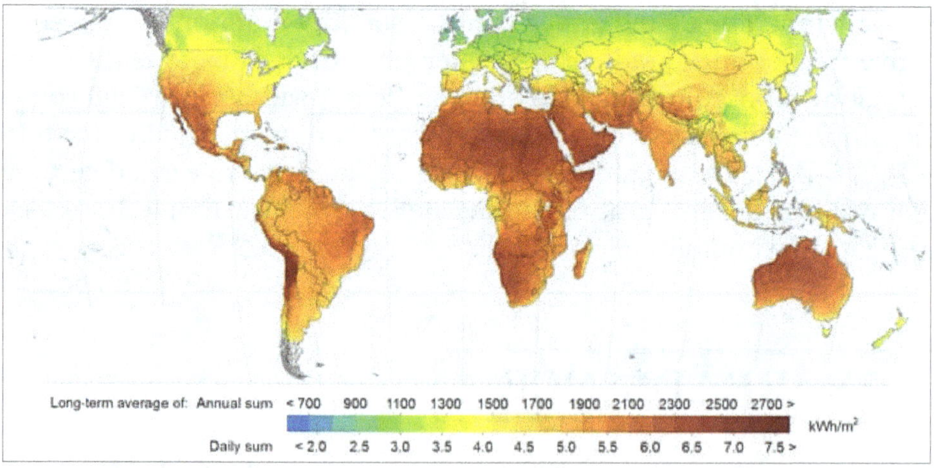

The image above shows the Earth's global solar irradiation.

Architects use solar irradiance to design passive heating and cooling systems for homes and buildings. Incorporating vertical windows on the side of the structure that faces the equator allows the sun's rays to enter for heating in the winter when the sun is low in the sky. Conversely, in the summer when the sun is higher in the sky during the day, less solar irradiation enters the structure, helping to keep it cool.

Solar irradiance also plays a key role in climate modeling and weather forecasting. Scientists combine insolation data with other quantitative measurements to study how climate and weather work, what affects them and to make predictive models to project future weather conditions and changes in climate.

References

- Aschwanden, Markus J. (2005). Physics of the Solar Corona: An Introduction with Problems and Solutions. Chichester, UK: Praxis Publishing. ISBN 978-3-540-22321-4
- what-is-solar-astronomy: actforlibraries.org, Retrieved 24 March, 2020
- Martins, F.; Depagne, E.; Russeil, D.; Mahy, L. (2013). "Evidence of quasi-chemically homogeneous evolution of massive stars up to solar metallicity". Astronomy & Astrophysics. 554: A23. arXiv:1304.3337. Bibcode:2013A&A...554A..23M. doi:10.1051/0004-6361/201321282
- Sun, entry: newworldencyclopedia.org, Retrieved 2 March, 2020
- Pascale Ehrenfreund; et al., eds. (2004). Astrobiology: future perspectives. Dordrecht [u.a.]: Kluwer Academic. ISBN 978-1-4020-2304-0. Retrieved 28 August, 2014
- What-are-sunspots, help: spaceweatherlive.com, Retrieved 16 June, 2020
- Solar Physics Laboratory, Code 671. "HINODE". NASA Goddard Space Flight Centre. Retrieved 17 January, 2014

Cosmology: A Comprehensive Study

Cosmology is the branch of astronomy which deals with the study of origin and evolution of universe. It is divided into physical cosmology, philosophical cosmology and religious cosmology. The Big Bang theory, dark matter and energy, are some of the concepts of cosmology. This chapter discusses all the concepts of cosmology in detail.

It is the branch of science that deals with the scientific study of the origins and evolution of the Universe and the nature of the Universe on its very largest scales.

In its earliest form physical cosmology was basically just celestial mechanics, the study of the 'heavens'. The Greek philosophers Aristarchus of Samos, Aristotle and Ptolemy proposed different theories for how the heavens work.

In particular, the 'earth-centric' Ptolemaic system with it's perfect circles and small 'epicycles' was the accepted theory to explain the motion of the heavenly bodies. In the picture above, the Earth is at the centre with the Sun, Moon and planets all revolving around Earth.

It was the best theory until Copernicus, Kepler and Galileo proposed a 'sun-centric' system in the 16th century. Although Greek, Indian and Muslim savants formulated the sun-centric theory centuries before Copernicus, his reiteration that the Sun, rather than the Earth, is at the center of the solar system is considered among the most important landmarks in the history of modern astronomy.

Newton's Cosmology

With Isaac Newton's 1687 publication of 'Principia Mathematica', the problem of the motion of the heavenly bodies was solved. Newton provided a physical mechanism for Kepler's laws of planetary motion. His law of universal gravitation resolved the anomalies caused by gravitational interaction between the planets in the previous systems.

Despite the words 'universal gravitation', Newton did not consider the implications of his theory on the Universe at large, although it implied that gravity works the same everywhere. The universal application of gravity was left to Albert Einstein, who formulated it more than two centuries later.

Heber D. Curtis, on the other hand, suggested that the observed spiral nebulae were star systems in their own right, or 'island universes'. In 1923-24 Edwin Hubble detected novae in the Andromeda galaxy and then showed its distance to be way beyond the Milky Way's boundaries. This settled the debate and it was accepted that the Milky Way was not all the universe there is. However, the cosmos was still thought to be static and unchanging.

Modern Cosmology

Subsequent modeling of the universe explored the possibility that the cosmological constant introduced by Einstein may result in an expanding universe, depending on its value. In 1929, Edwin Hubble's discovered the red shift of the light of distant galaxies, indicating that they move away from the Milky Way. Hence, the universe must be expanding.

Einstein's Cosmology

Scientific cosmology really began in 1917, when Albert Einstein's published the final modification to his theory of gravity in the paper 'Cosmological Considerations of the General Theory of Relativity'. This paper prompted early cosmologists such as Willem de Sitter, Karl Schwarzschild and Arthur Eddington to explore the astronomical consequences of the theory of relativity.

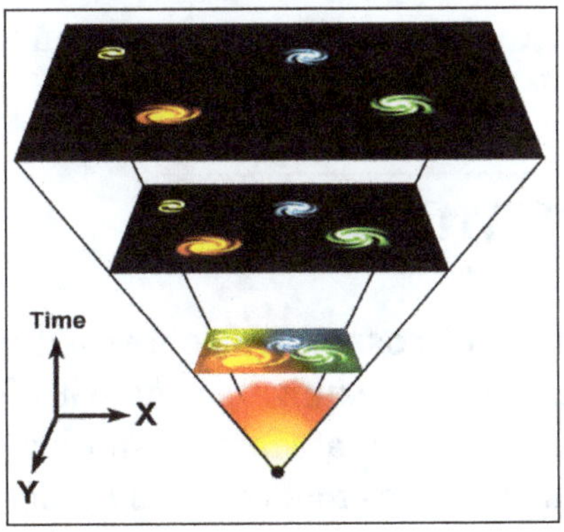

Mount Wilson astronomer Harlow Shapley championed the model of a cosmos made up of the Milky Way star system only – a static unchanging universe. Einstein also believed this and 'fudged' his equations of general relativity to represent this static state. He introduced a 'cosmological constant' that prevented the universe from contracting or expanding, which was what his original equations told him.

Einstein promptly repudiated his earlier 'fudge factor' and declared that the introduction of the cosmological constant was the "biggest blunder of his scientific career". His original 1917 equations support an expanding universe without the cosmological constant anyway.

In 1931 Georges Lemaitre, a Belgian priest, postulated that an expanding universe meant a creation event at some time in the past. He postulated that the Universe was created from a small 'cosmic egg' and set to expand until today. This theory was later labeled the 'big bang'.

There were many rival theories to the big bang. The 1964 discovery of the cosmic microwave background radiation by Arno Penzias and Robert Woodrow Wilson was a first step in ruling out many alternative physical cosmologies.

The Golden Age of Cosmology

Recent observations made by the COBE and WMAP satellites observing and accurately measuring this background radiation have effectively, transformed cosmology from a highly speculative science into a predictive science. This has led many to refer to modern times as the "golden age of cosmology".

Present observations match predictions made by a theory called Cosmic Inflation, first proposed by American physicist and cosmologist Alan Guth in 1981. It is a modification of the standard big bang theory that explains how the universe obtained the expansion and solves some problem areas of the big bang satisfactorily.

Since there are still problems with scientifically explaining how the universe came forth 'out of nothing', modern research are mainly concentrated on 'quantum cosmology'. This is a branch of 'quantum gravity', which strives to reconcile general relativity with quantum physics. No satisfactory reconciliation has been possible yet.

PHYSICAL COSMOLOGY

Physical cosmology is a branch of cosmology concerned with the studies of the largest-scale structures and dynamics of the universe and with fundamental questions about its origin, structure, evolution, and ultimate fate. Cosmology as a science originated with the Copernican principle, which implies that celestial bodies obey identical physical laws to those on Earth, and Newtonian mechanics, which first allowed those physical laws to be understood. Physical cosmology, as it is now understood, began with the development in 1915 of Albert Einstein's general theory of relativity, followed by major observational discoveries in the 1920s: first, Edwin Hubble discovered that the universe contains a huge number of external galaxies beyond the Milky Way; then, work by Vesto Slipher and others showed that the universe is expanding. These advances made it possible to speculate about the origin of the universe, and allowed the establishment of the Big Bang theory, by Georges Lemaître, as the leading cosmological model. A few researchers still advocate a handful of alternative cosmologies; however, most cosmologists agree that the Big Bang theory explains the observations better.

Dramatic advances in observational cosmology since the 1990s, including the cosmic microwave background, distant supernovae and galaxy redshift surveys, have led to the development of a standard model of cosmology. This model requires the universe to contain large amounts of dark matter and dark energy whose nature is currently not well understood, but the model gives detailed predictions that are in excellent agreement with many diverse observations.

Cosmology draws heavily on the work of many disparate areas of research in theoretical and applied physics. Areas relevant to cosmology include particle physics experiments and theory, theoretical and observational astrophysics, general relativity, quantum mechanics, and plasma physics.

Energy of the Cosmos

The lightest chemical elements, primarily hydrogen and helium, were created during the Big Bang through the process of nucleosynthesis. In a sequence of stellar nucleosynthesis reactions, smaller atomic nuclei are then combined into larger atomic nuclei, ultimately forming stable iron group elements such as iron and nickel, which have the highest nuclear binding energies. The net process results in a *later energy release*, meaning subsequent to the Big Bang. Such reactions of nuclear particles can lead to *sudden energy releases* from cataclysmic variable stars such as novae. Gravitational collapse of matter into black holes also powers the most energetic processes, generally seen in the nuclear regions of galaxies, forming *quasars* and *active galaxies*.

Cosmologists cannot explain all cosmic phenomena exactly, such as those related to the accelerating expansion of the universe, using conventional forms of energy. Instead, cosmologists propose a new form of energy called dark energy that permeates all space. One hypothesis is that dark energy is just the vacuum energy, a component of empty space that is associated with the virtual particles that exist due to the uncertainty principle.

There is no clear way to define the total energy in the universe using the most widely accepted theory of gravity, general relativity. Therefore, it remains controversial whether the total energy is conserved in an expanding universe. For instance, each photon that travels through intergalactic space loses energy due to the redshift effect. This energy is not obviously transferred to any other system, so seems to be permanently lost. On the other hand, some cosmologists insist that energy is conserved in some sense; this follows the law of conservation of energy.

Thermodynamics of the universe is a field of study that explores which form of energy dominates the cosmos – relativistic particles which are referred to as radiation, or non-relativistic particles referred to as matter. Relativistic particles are particles whose rest mass is zero or negligible compared to their kinetic energy, and so move at the speed of light or very close to it; non-relativistic particles have much higher rest mass than their energy and so move much slower than the speed of light.

As the universe expands, both matter and radiation in it become diluted. However, the energy densities of radiation and matter dilute at different rates. As a particular volume expands, mass energy density is changed only by the increase in volume, but the energy density of radiation is changed both by the increase in volume and by the increase in the wavelength of the photons that make it up. Thus the energy of radiation becomes a smaller part of the universe's total energy than that of matter as it expands. The very early universe is said to have been 'radiation dominated' and radiation controlled the deceleration of expansion. Later, as the average energy per photon becomes roughly 10 eV and lower, matter dictates the rate of deceleration and the universe is said to be 'matter dominated'. The intermediate case is not treated well analytically. As the expansion of the universe continues, matter dilutes even further and the cosmological constant becomes dominant, leading to an acceleration in the universe's expansion.

COSMOCHEMISTRY

Cosmochemistry or chemical cosmology is the study of the chemical composition of matter in the universe and the processes that led to those compositions. This is done primarily through the study of the chemical composition of meteorites and other physical samples. Given that the asteroid parent bodies of meteorites were some of the first solid material to condense from the early solar nebula, cosmochemists are generally, but not exclusively, concerned with the objects contained within the solar system.

In 1938, Swiss mineralogist Victor Goldschmidt and his colleagues compiled a list of what they called "cosmic abundances" based on their analysis of several terrestrial and meteorite samples. Goldschmidt justified the inclusion of meteorite composition data into his table by claiming that terrestrial rocks were subjected to a significant amount of chemical change due to the inherent processes of the Earth and the atmosphere. This meant that studying terrestrial rocks exclusively would not yield an accurate overall picture of the chemical composition of the cosmos. Therefore, Goldschmidt concluded that extraterrestrial material must also be included to produce more accurate and robust data. This research is considered to be the foundation of modern cosmochemistry.

During the 1950s and 1960s, cosmochemistry became more accepted as a science. Harold Urey, widely considered to be one of the fathers of cosmochemistry, engaged in research that eventually led to an understanding of the origin of the elements and the chemical abundance of stars. In 1956, Urey and his colleague, German scientist Hans Suess, published the first table of cosmic abundances to include isotopes based on meteorite analysis.

The continued refinement of analytical instrumentation throughout the 1960s, especially that of mass spectrometry, allowed cosmochemists to perform detailed analyses of the isotopic abundances of elements within meteorites. in 1960, John Reynolds determined, through the analysis of short-lived nuclides within meteorites, that the elements of the solar system were formed before the solar system itself which began to establish a timeline of the processes of the early solar system.

In October 2011, scientists reported that cosmic dust contains complex organic matter ("amorphous organic solids with a mixed aromatic-aliphatic structure") that could be created naturally, and rapidly, by stars.

On August 29, 2012, and in a world first, astronomers at Copenhagen University reported the detection of a specific sugar molecule, glycolaldehyde, in a distant star system. The molecule was found around the protostellar binary IRAS 16293-2422, which is located 400 light years from Earth. Glycolaldehyde is needed to form ribonucleic acid, or RNA, which is similar in function to DNA. This finding suggests that complex organic molecules may form in stellar systems prior to the formation of planets, eventually arriving on young planets early in their formation.

In September 2012, NASA scientists reported that polycyclic aromatic hydrocarbons (PAHs), subjected to interstellar medium (ISM) conditions, are transformed, through hydrogenation, oxygenation and hydroxylation, to more complex organics – "a step along the path toward amino acids and nucleotides, the raw materials of proteins and DNA, respectively". Further, as a result of these transformations, the PAHs lose their spectroscopic signature which could be one of the reasons "for the lack of PAH detection in interstellar ice grains, particularly the outer regions of cold, dense clouds or the upper molecular layers of protoplanetary disks."

Meteo Rites

Meteorites are one of the most important tools that cosmochemists have for studying the chemical nature of the solar system. Many meteorites come from material that is as old as the solar system itself, and thus provides scientists with a record from the early solar nebula. Carbonaceous chondrites are especially primitive; that is they have retained many of their chemical properties since their formation 4.56 billion years ago, and are therefore a major focus of cosmochemical investigations.

The most primitive meteorites also contain a small amount of material (< 0.1%) which is now recognized to be presolar grains that are older than the solar system itself, and which are derived directly from the remnants of the individual supernovae that supplied the dust from which the solar system formed. These grains are recognizable from their exotic chemistry which is alien to the solar system (such as matrixes of graphite, diamond, or silicon carbide). They also often have isotope ratios which are not those of the rest of the solar system (in particular, the Sun), and which differ from each other, indicating sources in a number of different explosive supernova events. Meteorites also may contain interstellar dust grains, which have collected from non-gaseous elements in the interstellar medium, as one type of composite cosmic dust ("stardust").

Recent findings by NASA, based on studies of meteorites found on Earth, suggests DNA and RNA components (adenine, guanine and related organic molecules), building blocks for life as we know it, may be formed extraterrestrially in outer space.

BIG BANG

The Big Bang theory is the prevailing cosmological model for the observable universe from the earliest known periods through its subsequent large-scale evolution. The model describes how the universe expanded from a very high-density and high-temperature state, and offers a comprehensive explanation for a broad range of phenomena, including the abundance of light elements, the cosmic microwave background (CMB), large scale structure and Hubble's law (the farther away galaxies are, the faster they are moving away from Earth). If the observed conditions are extrapolated backwards in time using the known laws of physics, the prediction is that just before a period of very high density there was a singularity which is typically associated with the Big Bang. Current knowledge is insufficient to determine if the singularity was primordial.

Since Georges Lemaître first noted in 1927 that an expanding universe could be traced back in time to an originating single point, scientists have built on his idea of cosmic expansion. The scientific community was once divided between supporters of two different theories, the Big Bang and the Steady State theory, but a wide range of empirical evidence has strongly favored the Big Bang which is now universally accepted. In 1929, from analysis of galactic redshifts, Edwin Hubble concluded that galaxies are drifting apart; this is important observational evidence for an expanding universe. In 1964, the cosmic microwave background radiation was discovered, which was crucial evidence in favor of the hot Big Bang model, since that theory predicted the existence of background radiation throughout the universe before it was discovered.

The known physical laws of nature can be used to calculate the characteristics of the universe in detail back in time to an initial state of extreme density and temperature. Detailed measurements of the expansion rate of the universe place the Big Bang at around 13.8 billion years ago, which is thus considered the age of the universe. After its initial expansion, the universe cooled sufficiently to allow the formation of subatomic particles, and later atoms. Giant clouds of these primordial elements (mostly hydrogen, with some helium and lithium) later coalesced through gravity, eventually forming early stars and galaxies, the descendants of which are visible today. Astronomers also observe the gravitational effects of dark matter surrounding galaxies. Most of the matter in the universe seems to be in the form of dark matter, and Big Bang theory and various observations indicate that it is not conventional baryonic matter (atoms). It is still not known exactly what dark matter is. More recently, measurements of the redshifts of supernovae indicate that the expansion of the universe is accelerating, an observation attributed to dark energy's existence.

In 1922, Russian mathematician Alexander Friedmann proposed on theoretical grounds that the universe is expanding, which was rederived independently and observationally confirmed soon afterwards by Belgian astronomer and Catholic priest Georges Lemaître in 1927. Lemaître also proposed what became known as the "Big Bang theory" of the creation of the universe, originally calling it the "hypothesis of the primeval atom". He presented his new idea that the universe is expanding and provided the first observational estimation of what is known as the Hubble constant. What later will be known as the "Big Bang theory" of the origin of the universe, he called his "hypothesis of the primeval atom" or the "Cosmic Egg".

American astronomer Edwin Hubble observed that the distances to faraway galaxies were strongly correlated with their redshifts. This was interpreted to mean that all distant galaxies and clusters are receding away from our vantage point with an apparent velocity proportional to their distance: that is, the farther they are, the faster they move away from us, regardless of direction. Assuming the Copernican principle (that the Earth is not the center of the universe), the only remaining interpretation is that all observable regions of the universe are receding from all others. Since we know that the distance between galaxies increases today, it must mean that in the past galaxies were closer together. The continuous expansion of the universe implies that the universe was denser and hotter in the past.

Large particle accelerators can replicate the conditions that prevailed after the early moments of the universe, resulting in confirmation and refinement of the details of the Big Bang model. However, these accelerators can only probe so far into high energy regimes. Consequently, the state of the universe in the earliest instants of the Big Bang expansion is still poorly understood and an area of open investigation and speculation.

The first subatomic particles to be formed included protons, neutrons, and electrons. Though simple atomic nuclei formed within the first three minutes after the Big Bang, thousands of years passed before the first electrically neutral atoms formed. The majority of atoms produced by the Big Bang were hydrogen, along with helium and traces of lithium. Giant clouds of these primordial elements later coalesced through gravity to form stars and galaxies, and the heavier elements were synthesized either within stars or during supernovae.

The Big Bang theory offers a comprehensive explanation for a broad range of observed phenomena, including the abundance of light elements, the CMB, large scale structure, and Hubble's Law. The framework for the Big Bang model relies on Albert Einstein's theory of general relativity and on simplifying assumptions such as homogeneity and isotropy of space. The governing equations were formulated by Alexander Friedmann, and similar solutions were worked on by Willem de Sitter. Since then, astrophysicists have incorporated observational and theoretical additions into the Big Bang model, and its parametrization as the Lambda-CDM model serves as the framework for current investigations of theoretical cosmology. The Lambda-CDM model is the current "standard model" of Big Bang cosmology, consensus is that it is the simplest model that can account for the various measurements and observations relevant to cosmology.

Timeline

Singularity

Extrapolation of the expansion of the universe backwards in time using general relativity yields an infinite density and temperature at a finite time in the past. This singularity indicates that general relativity is not an adequate description of the laws of physics in this regime. Models based on general relativity alone can not extrapolate toward the singularity beyond the end of the Planck epoch.

This primordial singularity is itself sometimes called "the Big Bang", but the term can also refer to a more generic early hot, dense phase of the universe. In either case, "the Big Bang" as an event is also colloquially referred to as the "birth" of our universe since it represents the point in history where the universe can be verified to have entered into a regime where the laws of physics as we understand them (specifically general relativity and the standard model of particle physics) work. Based on measurements of the expansion using Type Ia supernovae and measurements of temperature fluctuations in the cosmic microwave background, the time that has passed since that event — otherwise known as the "age of the universe" — is 13.799 ± 0.021 billion years. The agreement of independent measurements of this age supports the ΛCDM model that describes in detail the characteristics of the universe.

Despite being extremely dense at this time—far denser than is usually required to form a black hole—the universe did not re-collapse into a black hole. This may be explained by considering that commonly-used calculations and limits for gravitational collapse are usually based upon objects of relatively constant size, such as stars, and do not apply to rapidly expanding space such as the Big Bang.

Inflation and Baryogenesis

The earliest phases of the Big Bang are subject to much speculation. In the most common models the universe was filled homogeneously and isotropically with a very high energy density and huge temperatures and pressures and was very rapidly expanding and cooling. Approximately 10–37 seconds into the expansion, a phase transition caused a cosmic inflation, during which the universe grew exponentially during which time density fluctuations that occurred because of the uncertainty principle were amplified into the seeds that would later form the large-scale structure of the universe. After inflation stopped, reheating occurred until the universe obtained the temperatures required for the production of a quark–gluon plasma as well as all other elementary particles. Temperatures were so high that the random motions of particles were at relativistic speeds, and particle–antiparticle pairs of all kinds were being continuously created and destroyed in collisions. At some point, an unknown reaction called baryogenesis violated the conservation of baryon number, leading to a very small excess of quarks and leptons over antiquarks and antileptons—of the order of one part in 30 million. This resulted in the predominance of matter over antimatter in the present universe.

Cooling

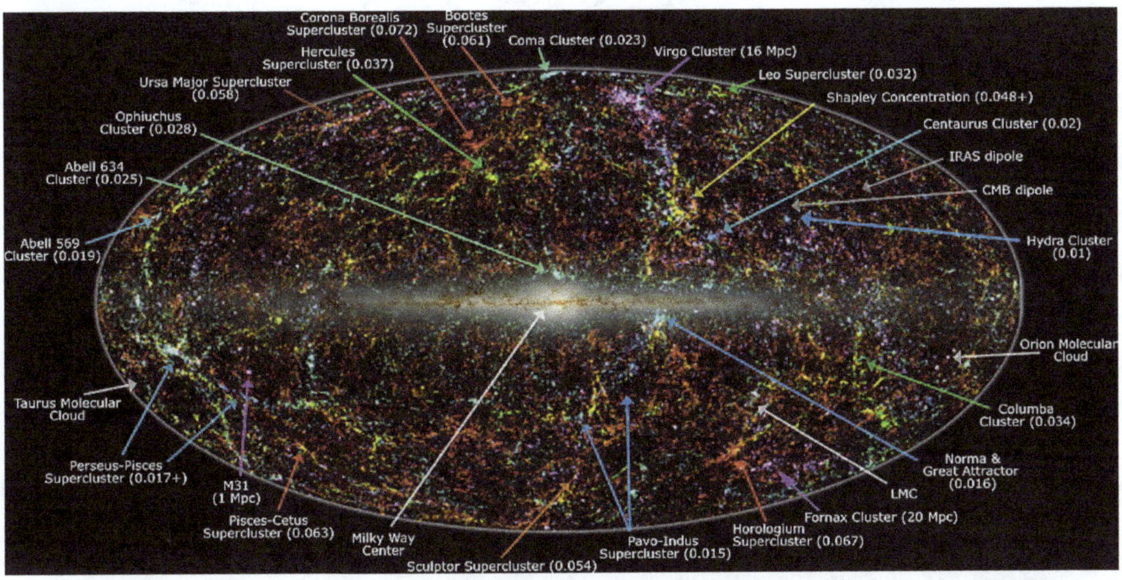

Panoramic view of the entire near-infrared sky reveals the distribution of galaxies beyond the Milky Way. Galaxies are color-coded by redshift.

The universe continued to decrease in density and fall in temperature, hence the typical energy of each particle was decreasing. Symmetry breaking phase transitions put the fundamental forces of physics and the parameters of elementary particles into their present form. After about 10^{-11} seconds, the picture becomes less speculative, since particle energies drop to values that can be attained in particle accelerators. At about 10^{-6} seconds, quarks and gluons combined to form baryons such as protons and neutrons.

The small excess of quarks over antiquarks led to a small excess of baryons over antibaryons. The temperature was now no longer high enough to create new proton–antiproton pairs (similarly for neutrons–antineutrons), so a mass annihilation immediately followed, leaving just one in 10^{10} of the original protons and neutrons, and none of their antiparticles. A similar process happened at about 1 second for electrons and positrons. After these annihilations, the remaining protons, neutrons and electrons were no longer moving relativistically and the energy density of the universe was dominated by photons (with a minor contribution from neutrinos).

A few minutes into the expansion, when the temperature was about a billion (one thousand million) kelvin and the density was about that of air, neutrons combined with protons to form the universe's deuterium and helium nuclei in a process called Big Bang nucleosynthesis. Most protons remained uncombined as hydrogen nuclei.

As the universe cooled, the rest mass energy density of matter came to gravitationally dominate that of the photon radiation. After about 379,000 years, the electrons and nuclei combined into atoms (mostly hydrogen); hence the radiation decoupled from matter and continued through space largely unimpeded. This relic radiation is known as the cosmic microwave background radiation. The chemistry of life may have begun shortly after the Big Bang, 13.8 billion years ago, during a habitable epoch when the universe was only 10–17 million years old.

Lambda-CDM Model

The ΛCDM (Lambda cold dark matter) or Lambda-CDM model is a parametrization of the Big Bang cosmological model in which the universe contains three major components: first, a cosmological constant denoted by Lambda (Greek Λ) and associated with dark energy; second, the postulated cold dark matter (abbreviated CDM); and third, ordinary matter. It is frequently referred to as the standard model of Big Bang cosmology because it is the simplest model that provides a reasonably good account of the following properties of the cosmos:

- The existence and structure of the cosmic microwave background.
- The large-scale structure in the distribution of galaxies.
- The abundances of hydrogen (including deuterium), helium, and lithium.
- The accelerating expansion of the universe observed in the light from distant galaxies and supernovae.

The model assumes that general relativity is the correct theory of gravity on cosmological scales. It emerged in the late 1990s as a concordance cosmology, after a period of time when disparate observed properties of the universe appeared mutually inconsistent, and there was no consensus on the makeup of the energy density of the universe.

The ΛCDM model can be extended by adding cosmological inflation, quintessence and other elements that are current areas of speculation and research in cosmology.

Some alternative models challenge the assumptions of the ΛCDM model. Examples of these are modified Newtonian dynamics, entropic gravity, modified gravity, theories of large-scale variations in the matter density of the universe, bimetric gravity, and scale invariance of empty space.

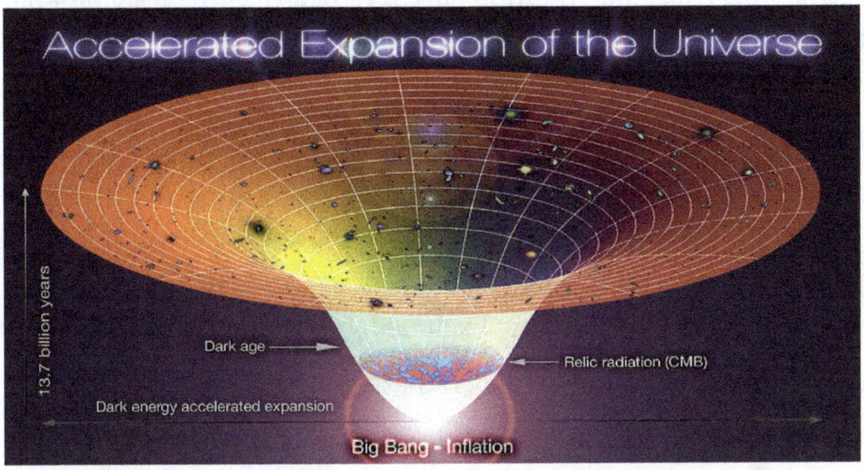

Lambda-CDM, accelerated expansion of the universe. The time-line in this schematic diagram extends from the Big Bang/inflation era 13.7 Byr ago to the present cosmological time.

Most modern cosmological models are based on the cosmological principle, which states that our observational location in the universe is not unusual or special; on a large-enough scale, the universe looks the same in all directions (isotropy) and from every location (homogeneity).

The model includes an expansion of metric space that is well documented both as the red shift of prominent spectral absorption or emission lines in the light from distant galaxies and as the time dilation in the light decay of supernova luminosity curves. Both effects are attributed to a Doppler shift in electromagnetic radiation as it travels across expanding space. Although this expansion increases the distance between objects that are not under shared gravitational influence, it does not increase the size of the objects (e.g. galaxies) in space. It also allows for distant galaxies to recede from each other at speeds greater than the speed of light; local expansion is less than the speed of light, but expansion summed across great distances can collectively exceed the speed of light.

The letter Λ (lambda) represents the cosmological constant, which is currently associated with a vacuum energy or dark energy in empty space that is used to explain the contemporary accelerating expansion of space against the attractive effects of gravity. A cosmological constant has negative pressure, $p = -\rho c^2$, which contributes to the stress-energy tensor that, according to the general theory of relativity, causes accelerating expansion. The fraction of the total energy density of our (flat or almost flat) universe that is dark energy, Ω_Λ, is estimated to be 0.669 ± 0.038 based on the 2018 Dark Energy Survey results using Type Ia Supernovae or 0.6847 ± 0.0073 based on the 2018 release of Planck satellite data, or more than 68.3% (2018 estimate) of the mass-energy density of the universe.

Dark matter is postulated in order to account for gravitational effects observed in very large-scale structures (the "flat" rotation curves of galaxies; the gravitational lensing of light by galaxy clusters; and enhanced clustering of galaxies) that cannot be accounted for by the quantity of observed matter.

Cold dark matter as currently hypothesized is:

- Non-baryonic: It consists of matter other than protons and neutrons (and electrons, by convention, although electrons are not baryons).
- Cold: Its velocity is far less than the speed of light at the epoch of radiation-matter equality (thus neutrinos are excluded, being non-baryonic but not cold).
- Dissipationless: It cannot cool by radiating photons.
- Collisionless: The dark matter particles interact with each other and other particles only through gravity and possibly the weak force.

Dark matter constitutes about 26.8% (2013 estimate) of the mass-energy density of the universe. The remaining 4.8% (2013 estimate) comprises all ordinary matter observed as atoms, chemical elements, gas and plasma, the stuff of which visible planets, stars and galaxies are made. The great majority of ordinary matter in the universe is unseen, since visible stars and gas inside galaxies and clusters account for less than 10% of the ordinary matter contribution to the mass-energy density of the universe.

Also, the energy density includes a very small fraction (~ 0.01%) in cosmic microwave background radiation, and not more than 0.5% in relic neutrinos. Although very small today, these were much more important in the distant past, dominating the matter at redshift > 3200.

The model includes a single originating event, the "Big Bang", which was not an explosion but the abrupt appearance of expanding space-time containing radiation at temperatures of around 10^{15} K. This was immediately (within 10^{-29} seconds) followed by an exponential expansion of space by a scale multiplier of 10^{27} or more, known as cosmic inflation. The early universe remained hot (above 10,000 K) for several hundred thousand years, a state that is detectable as a residual cosmic microwave background, or CMB, a very low energy radiation emanating from all parts of the sky. The "Big Bang" scenario, with cosmic inflation and standard particle physics, is the only current cosmological model consistent with the observed continuing expansion of space, the observed distribution of lighter elements in the universe (hydrogen, helium, and lithium), and the spatial texture of minute irregularities (anisotropies) in the CMB radiation. Cosmic inflation also addresses the "horizon problem" in the CMB; indeed, it seems likely that the universe is larger than the observable particle horizon.

The model uses the Friedmann–Lemaître–Robertson–Walker metric, the Friedmann equations and the cosmological equations of state to describe the observable universe from right after the inflationary epoch to present and future.

Cosmic Expansion History

The expansion of the universe is parameterized by a dimensionless scale factor $a = a(t)$ (with time t counted from the birth of the universe), defined relative to the present day, so $a_0 = a(t_0) = 1$; the usual convention in cosmology is that subscript 0 denotes present-day values, so t_0 is the current age of the universe. The scale factor is related to the observed redshift z of the light emitted at time t_{em} by:

$$a(t_{\text{em}}) = \frac{1}{1+z}.$$

The expansion rate is described by the time-dependent Hubble parameter, $H(t)$, defined as:

$$H(t) \equiv \frac{\dot{a}}{a},$$

where $\dot{a}$ is the time-derivative of the scale factor. The first Friedmann equation gives the expansion rate in terms of the matter+radiation density ρ, the curvature k, and the cosmological constant Λ:

$$H^2 = \left(\frac{\dot{a}}{a}\right)^2 = \frac{8\pi G}{3}\rho - \frac{kc^2}{a^2} + \frac{\Lambda c^2}{3},$$

where as usual c is the speed of light and G is the gravitational constant. A critical density ρ_{crit} is the present-day density, which gives zero curvature k, assuming the cosmological constant Λ is zero, regardless of its actual value. Substituting these conditions to the Friedmann equation gives:

$$\rho_{\text{crit}} = \frac{3H_0^2}{8\pi G} = 1.878\,47(23) \times 10^{-26}\, h^2 \text{ kg m}^{-3},$$

where $h \equiv H_0 / (100 \text{ km s}^{-1} \text{ Mpc}^{-1})$ is the reduced Hubble constant. If the cosmological constant were actually zero, the critical density would also mark the dividing line between eventual recollapse of the universe to a Big Crunch, or unlimited expansion. For the Lambda-CDM model with a positive cosmological constant (as observed), the universe is predicted to expand forever regardless of whether the total density is slightly above or below the critical density; though other outcomes are possible in extended models where the dark energy is not constant but actually time-dependent.

It is standard to define the present-day density parameter Ω_x for various species as the dimensionless ratio:

$$\Omega_x \equiv \frac{\rho_x(t=t_0)}{\rho_{\text{crit}}} = \frac{8\pi G\rho_x(t=t_0)}{3H_0^2}$$

where the subscript x is one of b for baryons, c for cold dark matter, rad for radiation (photons plus relativistic neutrinos), and DE or Λ for dark energy.

Since the densities of various species scale as different powers of a, e.g. a^{-3} for matter etc., the Friedmann equation can be conveniently rewritten in terms of the various density parameters as:

$$H(a) \equiv \frac{\dot{a}}{a} = H_0\sqrt{(\Omega_c + \Omega_b)a^{-3} + \Omega_{\text{rad}}a^{-4} + \Omega_k a^{-2} + \Omega_{DE}a^{-3(1+w)}}$$

where w is the equation of state of dark energy, and assuming negligible neutrino mass (significant neutrino mass requires a more complex equation). The various Ω parameters add up to 1 by construction. In the general case this is integrated by computer to give the expansion history $a(t)$ and also observable distance-redshift relations for any chosen values of the cosmological parameters, which can then be compared with observations such as supernovae and baryon acoustic oscillations.

In the minimal 6-parameter Lambda-CDM model, it is assumed that curvature Ω_k is zero and $w = -1$, so this simplifies to:

$$H(a) = H_0\sqrt{\Omega_m a^{-3} + \Omega_{\text{rad}}a^{-4} + \Omega_\Lambda}.$$

Observations show that the radiation density is very small today, $\Omega_{\text{rad}} \sim 10^{-4}$; if this term is neglected the above has an analytic solution:

$$a(t) = (\Omega_m / \Omega_\Lambda)^{1/3}\sinh^{2/3}(t / t_\Lambda)$$

where $t_\Lambda \equiv 2/(3H_0\sqrt{\Omega_\Lambda})$; this is fairly accurate for $a > 0.01 a > 0.01 or t > 10 t > 10$ million years. Solving for $a(t) = 1$ gives the present age of the universe t_0 in terms of the other parameters.

It follows that the transition from decelerating to accelerating expansion (the second derivative $\ddot{a}$ crossing zero) occurred when:

$$a = (\Omega_m / 2\Omega_\Lambda)^{1/3}$$

which evaluates to $a \sim 0.6$ or $z \sim 0.66$ for the best-fit parameters estimated from the Planck spacecraft.

Successes

In addition to explaining pre-2000 observations, the model has made a number of successful predictions: notably the existence of the baryon acoustic oscillation feature, discovered in 2005 in the predicted location; and the statistics of weak gravitational lensing, first observed in 2000 by several teams. The polarization of the CMB, discovered in 2002 by DASI is now a dramatic success: in the 2015 Planck data release, there are seven observed peaks in the temperature (TT) power spectrum, six peaks in the temperature-polarization (TE) cross spectrum, and five peaks in the polarization (EE) spectrum. The six free parameters can be well constrained by the TT spectrum alone, and then the TE and EE spectra can be predicted theoretically to few-percent precision with no further adjustments allowed: comparison of theory and observations shows an excellent match.

Challenges

Extensive searches for dark matter particles have so far shown no well-agreed detection; the dark energy may be almost impossible to detect in a laboratory, and its value is unnaturally small compared to naive theoretical predictions.

Comparison of the model with observations is very successful on large scales (larger than galaxies, up to the observable horizon), but may have some problems on sub-galaxy scales, possibly predicting too many dwarf galaxies and too much dark matter in the innermost regions of galaxies. This problem is called the "small scale crisis". These small scales are harder to resolve in computer simulations, so it is not yet clear whether the problem is the simulations, non-standard properties of dark matter, or a more radical error in the model.

It has been argued that the ΛCDM model is built upon a foundation of conventionalist stratagems, rendering it unfalsifiable in the sense defined by Karl Popper.

Parameters

Planck Collaboration Cosmological parameters			
	Description	Symbol	Value
Independent parameters	Physical baryon density parameter	$\Omega_b h^2$	0.02230±0.00014
	Physical dark matter density parameter	$\Omega_c h^2$	0.1188±0.0010
	Age of the universe	t_0	13.799±0.021 × 10^9 years
	Scalar spectral index	n_s	0.9667±0.0040
	Curvature fluctuation amplitude, k_0 = 0.002 Mpc^{-1}	Δ_R^2	$2.441^{+0.088}_{-0.092} \times 10^{-9}$
	Reionization optical depth	τ	0.066±0.012

Fixed parameters	Total density parameter	Ω_{tot}	1
	Equation of state of dark energy	w	−1
	Tensor/scalar ratio	r	0
	Running of spectral index	$dn_s / d\ln k$	0
	Sum of three neutrino masses	$\sum m_\nu$	0.06 eV/c^2
	Effective number of relativistic degrees of freedom	N_{eff}	3.046
Calculated values	Hubble constant	H_0	67.74±0.46 km s^{-1} Mpc^{-1}
	Baryon density parameter	Ω_b	0.0486±0.0010
	Dark matter density parameter	Ω_c	0.2589±0.0057
	Matter density parameter	Ω_m	0.3089±0.0062
	Dark energy density parameter	Ω_Λ	0.6911±0.0062
	Critical density	ρ_{crit}	(8.62±0.12)×10^{-27} kg/m^3
	The present root-mean-square matter fluctuation averaged over a sphere of radius 8h^{-1} Mpc	σ_8	0.8159±0.0086
	Redshift at decoupling	z_*	1089.90±0.23
	Age at decoupling	t_*	377700±3200 years
	Redshift of reionization (with uniform prior)	z_{re}	$8.5^{+1.0}_{-1.1}$

The simple ΛCDM model is based on six parameters: physical baryon density parameter; physical dark matter density parameter; the age of the universe; scalar spectral index; curvature fluctuation amplitude; and reionization optical depth. In accordance with Occam's razor, six is the smallest number of parameters needed to give an acceptable fit to current observations; other possible parameters are fixed at "natural" values, e.g. total density parameter = 1.00, dark energy equation of state = −1.

The values of these six parameters are mostly not predicted by current theory (though, ideally, they may be related by a future "Theory of Everything"), except that most versions of cosmic inflation predict the scalar spectral index should be slightly smaller than 1, consistent with the estimated value 0.96. The parameter values, and uncertainties, are estimated using large computer searches to locate the region of parameter space providing an acceptable match to cosmological observations. From these six parameters, the other model values, such as the Hubble constant and the dark energy density, can be readily calculated.

Commonly, the set of observations fitted includes the cosmic microwave background anisotropy, the brightness/redshift relation for supernovae, and large-scale galaxy clustering including the baryon acoustic oscillation feature. Other observations, such as the Hubble constant, the abundance of galaxy clusters, weak gravitational lensing and globular cluster ages, are generally consistent with these, providing a check of the model, but are less precisely measured at present.

Parameter values listed below are from the Planck Collaboration Cosmological parameters 68% confidence limits for the base ΛCDM model from Planck CMB power spectra, in combination with lensing reconstruction and external data (BAO + JLA + H_0).

1. The "physical baryon density parameter" $\Omega_b h^2$ is the "baryon density parameter" Ω_b multiplied by the square of the reduced Hubble constant $h = H_0 / (100 \text{ km s}^{-1} \text{ Mpc}^{-1})$. Likewise for the difference between "physical dark matter density parameter" and "dark matter density parameter".
2. A density $\rho_x = \Omega_x \rho_{crit}$ is expressed in terms of the critical density ρ_{crit}, which is the total density of matter/energy needed for the universe to be spatially flat. Measurements indicate that the actual total density ρ_{tot} is very close if not equal to this value.
3. This is the minimal value allowed by solar and terrestrial neutrino oscillation experiments.
4. From the Standard Model of particle physics.
5. Calculated from $\Omega_b h^2$ and $h = H_0 / (100 \text{ km s}^{-1} \text{ Mpc}^{-1})$.
6. Calculated from $\Omega_c h^2$ and $h = H_0 / (100 \text{ km s}^{-1} \text{ Mpc}^{-1})$.
7. Calculated from $h = H_0 / (100 \text{ km s}^{-1} \text{ Mpc}^{-1})$ per $\rho_{crit} = 1.87847 \times 10^{-26} h^2 \text{ kg m}^{-3}$.

Missing Baryon Problem

Massimo Persic and Paolo Salucci firstly estimated the baryonic density today present in ellipticals, spirals, groups and clusters of galaxies. They performed an integration of the baryonic mass-to-light ratio over luminosity (in the following M_b / L), weighted with the luminosity function $\phi(L)$ over the previously mentioned classes of astrophysical objects:

$$\rho_b = \sum \int L\phi(L) \frac{M_b}{L} dL.$$

The result was:

$$\Omega_b = \Omega_* + \Omega_{gas} = 2.2 \times 10^{-3} + 1.5 \times 10^{-3} h^{-1.3} \simeq 0.003$$

where $h \simeq 0.72$.

Note that this value is much lower than the prediction of standard cosmic nucleosynthesis $\Omega_b \simeq 0.0486$, so that stars and gas in galaxies and in galaxy groups and clusters account for less than 10% of the primordially synthesized baryons. This issue is known as the problem of the "missing baryons".

Extended Models

Extended model parameters		
Description	Symbol	Value
Total density parameter	Ω_{tot}	$1.0023\,^{+0.0056}_{-0.0054}$
Equation of state of dark energy	w	−0.980±0.053
Tensor-to-scalar ratio	r	< 0.11, k_0 = 0.002 Mpc^{-1} (2σ)
Running of the spectral index	$dn_s / d\ln k$	−0.022±0.020, k_0 = 0.002 Mpc^{-1}
Sum of three neutrino masses	$\sum m_\nu$	< 0.58 eV/c^2 (2σ)
Physical neutrino density parameter	$\Omega_\nu h^2$	< 0.0062

Extended models allow one or more of the "fixed" parameters above to vary, in addition to the basic six; so these models join smoothly to the basic six-parameter model in the limit that the additional parameter(s) approach the default values. For example, possible extensions of the simplest ΛCDM model allow for spatial curvature (Ω_{tot} may be different from 1); or quintessence rather than a cosmological constant where the equation of state of dark energy is allowed to differ from −1. Cosmic inflation predicts tensor fluctuations (gravitational waves). Their amplitude is parameterized by the tensor-to-scalar ratio (denoted *r*), which is determined by the unknown energy scale of inflation. Other modifications allow hot dark matter in the form of neutrinos more massive than the minimal value, or a running spectral index; the latter is generally not favoured by simple cosmic inflation models.

Allowing additional variable parameter(s) will generally increase the uncertainties in the standard six parameters quoted above, and may also shift the central values slightly. The Table above shows results for each of the possible "6+1" scenarios with one additional variable parameter; this indicates that, as of 2015, there is no convincing evidence that any additional parameter is different from its default value.

Some researchers have suggested that there is a running spectral index, but no statistically significant study has revealed one. Theoretical expectations suggest that the tensor-to-scalar ratio r should be between 0 and 0.3, and the latest results are now within those limits.

Reionization

The Epoch of Reionization (EOR) refers to the period in the history of the universe during which the predominantly neutral intergalactic medium was ionized by the emergence of the first luminous sources. These sources may have been stars, galaxies, quasars, or some combination of the above. By studying reionization, we can learn a great deal about the process of structure formation in the universe, and find the evolutionary links between the remarkably smooth matter distribution at early times revealed by CMB studies, and the highly structured universe of galaxies and clusters of galaxies at redshifts of 6 and below. The Low Frequency Demonstrator is designed to provide detailed information on conditions in the intergalactic medium during and immediately preceding the EOR.

After the Big Bang, the Universe was a hot, but quickly cooling soup of fundamental particles. After a few hundred thousand years, things cooled enough so that protons and electrons could combine to form neutral hydrogen. This was a rather sudden event, and allowed the thermal glow of the fireball plasma, as it existed immediately before the hydrogen formation event, to radiate throughout the universe unimpeded by constant interactions with the charged particles of the now-absent plasma. This glow, redshifted by a factor of 1100 or so, is what we now observe as the Cosmic Microwave Background (CMB) in all directions. The CMB carries a frozen imprint of the density fluctuations in the early Universe, the study of which, by the observational cosmology community, is intense and sustained.

After the Universe became neutral, it became unobservable across much of the electromagnetic spectrum. Any short wavelength radiation that might have been emitted was quickly absorbed by the atomic gas, and a long interval known as the Dark Ages began. Slowly, gravitational collapse of overdense regions, the same regions we can see in the CMB imprint from earlier times, led to the formation of more and more pronounced structure in the neutral medium, and eventually the first stars, galaxies and quasars started to form. The exact mechanism and nature of this formation, poorly constrained by observation, is a topic of much research and great importance. We know what the Universe looked like at the time of the CMB, and we know what it looks like now, but how did it get from one to the other?

As the collapse of structures proceeded, temperature variations developed. Gradually, energetic radiation emitted by the first sources caused local heating, and then ionization of the hydrogen in the Universe. It will have started with "bubbles" of ionized plasma surrounding the most energetic sources. As the bubbles grew and became more numerous, they started to overlap, and more and more of the neutral medium became exposed to the harsh ionizing radiation, which travels unimpeded through ionized regions. The final phase of reionization of the Universe may have occurred swiftly. As soon as the bulk of the Universe was reionized, light at many wavelengths could escape from the early galaxies and quasars, revealing the distant Universe that we see today with optical and infrared telescopes.

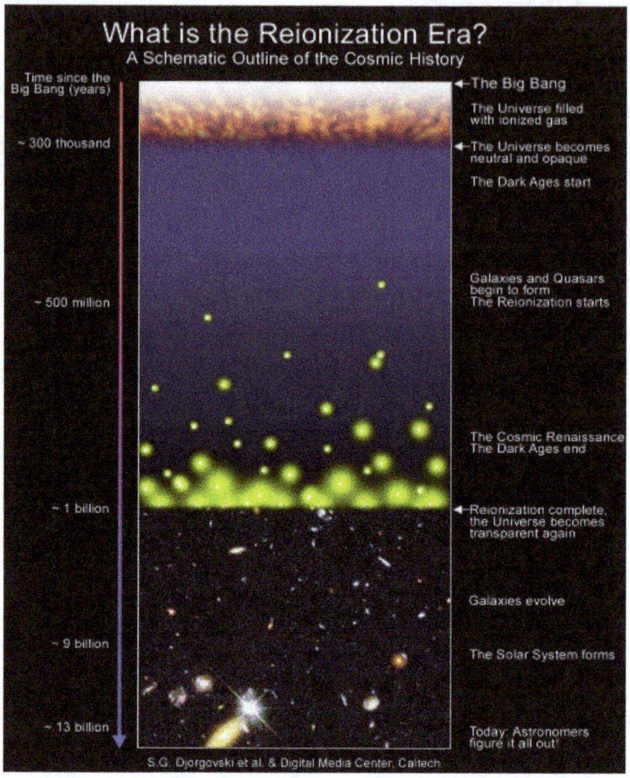

Graphical representation of the history of the universe.

Reionization was complete about 1 billion years after the Big Bang, corresponding to a redshift of about 6.5. Before that time, observations rapidly become more difficult. By and large, one must hope to find isolated, very luminous objects whose radiation in one form or another manages to reach us through the increasingly neutral medium. Perhaps the best hope for a more general and comprehensive probe of these early epochs is the 21cm hyperfine transition line of neutral hydrogen, redshifted to frequencies below 200 MHz. Sensitive observations of emission and absorption in this line can probe deeply into the reheating and reionization epochs, and give us a detailed view of the density, temperature and velocity field of the material. We would get a view, not just of isolated luminous objects and the material which happens to lie in front of them, but of large volumes of the Universe at the target redshifts. Such a view would yield a treasure trove of information from which to deduce the early history of structure formation, and the origin of the stars, galaxies, clusters and quasars that we see today.

BIG BANG NUCLEOSYNTHESIS

In physical cosmology, Big Bang nucleosynthesis (abbreviated BBN, also known as primordial nucleosynthesis, arch(a)eonucleosynthesis, archonucleosynthesis, protonucleosynthesis and pal(a)eonucleosynthesis) refers to the production of nuclei other than those of the lightest isotope of hydrogen (hydrogen-1, ^1H, having a single proton as a nucleus) during the early phases of the Universe. Primordial nucleosynthesis is believed by most cosmologists to have taken place in the interval from roughly 10 seconds to 20 minutes after the Big Bang, and is calculated to be responsible for the formation of most of the universe's helium as the isotope helium-4 (^4He), along with small amounts of the hydrogen isotope deuterium (^2H or D), the helium isotope helium-3 (^3He) and a very small amount of the lithium isotope lithium-7 (^7Li). In addition to these stable nuclei, two unstable or radioactive isotopes were also produced: the heavy hydrogen isotope tritium (^3H or T); and the beryllium isotope beryllium-7 (^7Be); but these unstable isotopes later decayed into ^3He and ^7Li, as above.

Essentially all of the elements that are heavier than lithium were created much later, by stellar nucleosynthesis in evolving and exploding stars.

Characteristics

There are several important characteristics of Big Bang nucleosynthesis (BBN):

- The initial conditions (neutron-proton ratio) were set in the first second after the Big Bang.
- The universe was very close to homogeneous at this time, and strongly radiation-dominated.
- The fusion of nuclei occurred between roughly 10 seconds to 20 minutes after the Big Bang; this corresponds to the temperature range when the universe was cool enough for deuterium to survive, but hot and dense enough for fusion reactions to occur at a significant rate.
- It was widespread, encompassing the entire observable universe.

The key parameter which allows one to calculate the effects of BBN is the baryon/photon number ratio, which is a small number of order 6×10^{-10}. This parameter corresponds to the baryon density and controls the rate at which nucleons collide and react; from this it is possible to calculate element abundances after nucleosynthesis ends. Although the baryon per photon ratio is important in determining element abundances, the precise value makes little difference to the overall picture. Without major changes to the Big Bang theory itself, BBN will result in mass abundances of about 75% of hydrogen-1, about 25% helium-4, about 0.01% of deuterium and helium-3, trace amounts (on the order of 10^{-10}) of lithium, and negligible heavier elements. That the observed abundances in the universe are generally consistent with these abundance numbers is considered strong evidence for the Big Bang theory.

In this field, for historical reasons it is customary to quote the helium-4 fraction by mass, symbol Y, so that 25% helium-4 means that helium-4 atoms account for 25% of the mass, but less than 8% of the nuclei would be helium-4 nuclei. Other (trace) nuclei are usually expressed as number ratios to hydrogen.

The first detailed calculations of the primordial isotopic abundances came in 1966 and have been refined over the years using updated estimates of the input nuclear reaction rates. The first systematic Monte Carlo study of how nuclear reaction rate uncertainties, over the relevant temperature range, impact isotope predictions was carried out in 1993.

Important Parameters

The creation of light elements during BBN was dependent on a number of parameters; among those was the neutron-proton ratio (calculable from Standard Model physics) and the baryon-photon ratio.

Neutron–proton Ratio

The neutron-proton ratio was set by Standard Model physics before the nucleosynthesis era, essentially within the first 1-second after the Big Bang. Neutrons can react with positrons or electron neutrinos to create protons and other products in one of the following reactions:

$$n + e^{+} \rightleftharpoons \bar{\nu}_e + p$$
$$n + \nu_e \rightleftharpoons p + e^{-}$$

At times much earlier than 1 sec, these reactions were fast and maintained the n/p ratio close to 1:1. As the temperature dropped, the equilibrium shifted in favour of protons due to their slightly lower mass, and the n/p ratio smoothly decreased. These reactions continued until the decreasing temperature and density caused the reactions to become too slow, which occurred at about T = 0.7 MeV (time around 1 second) and is called the freeze out temperature. At freeze out, the neutron-proton ratio was about 1/6. However, free neutrons are unstable with a mean life of 880 sec; some neutrons decayed in the next few minutes before fusing into any nucleus, so the ratio of total neutrons to protons after nucleosynthesis ends is about 1/7. Almost all neutrons that fused instead of decaying ended up combined into helium-4, due to the fact that helium-4 has the highest binding energy per nucleon among light elements. This predicts that about 8% of all atoms should be helium-4, leading to a mass fraction of helium-4 of about 25%, which is in line with observations. Small traces of deuterium and helium-3 remained as there was insufficient time and density for them to react and form helium-4.

Baryon–photon Ratio

The baryon–photon ratio, η, is the key parameter determining the abundances of light elements after nucleosynthesis ends. Baryons and light elements can fuse in the following main reactions:

$$p + n -> {}^2H + \gamma$$
$$p + {}^2H -> {}^3He + \gamma$$
$${}^2H + {}^2H -> {}^3He + n$$
$${}^2H + {}^2H -> {}^3H + p$$
$${}^3He + {}^2H -> {}^4He + p$$
$${}^3H + {}^2H -> {}^4He + n$$

along with some other low-probability reactions leading to ^{7}Li or ^{7}Be. (An important feature is that there are no stable nuclei with mass 5 or 8, which implies that reactions adding one baryon to ^{4}He, or fusing two ^{4}He, do not occur). Most fusion chains during BBN ultimately terminate in ^{4}He (helium-4), while "incomplete" reaction chains lead to small amounts of left-over ^{2}H or ^{3}He; the amount of these decreases with increasing baryon-photon ratio. That is, the larger the baryon-photon ratio the more reactions there will be and the more efficiently deuterium will be eventually transformed into helium-4. This result makes deuterium a very useful tool in measuring the baryon-to-photon ratio.

Sequence

Big Bang nucleosynthesis began roughly 10 seconds after the big bang, when the universe had cooled sufficiently to allow deuterium nuclei to survive disruption by high-energy photons. (Note that the neutron-proton freeze-out time was earlier). This time is essentially independent of dark matter content, since the universe was highly radiation dominated until much later, and this dominant component controls the temperature/time relation. At this time there were about six protons for every neutron, but a small fraction of the neutrons decay before fusing in the next few hundred seconds, so at the end of nucleosynthesis there are about seven protons to every neutron, and almost all the neutrons are in Helium-4 nuclei.

One feature of BBN is that the physical laws and constants that govern the behavior of matter at these energies are very well understood, and hence BBN lacks some of the speculative uncertainties that characterize earlier periods in the life of the universe. Another feature is that the process of nucleosynthesis is determined by conditions at the start of this phase of the life of the universe, and proceeds independently of what happened before.

As the universe expands, it cools. Free neutrons are less stable than helium nuclei, and the protons and neutrons have a strong tendency to form helium-4. However, forming helium-4 requires the intermediate step of forming deuterium. Before nucleosynthesis began, the temperature was high enough for many photons to have energy greater than the binding energy of deuterium; therefore any deuterium that was formed was immediately destroyed (a situation known as the deuterium bottleneck). Hence, the formation of helium-4 is delayed until the universe became cool enough for deuterium to survive (at about T = 0.1 MeV); after which there was a sudden burst of element formation. However, very shortly thereafter, around twenty minutes after the Big Bang, the temperature and density became too low for any significant fusion to occur. At this point, the elemental abundances were nearly fixed, and the only changes were the result of the radioactive decay of the two major unstable products of BBN, tritium and beryllium-7.

Heavy Elements

Big Bang nucleosynthesis produced very few nuclei of elements heavier than lithium due to a bottleneck: the absence of a stable nucleus with 8 or 5 nucleons. This deficit of larger atoms also limited the amounts of lithium-7 produced during BBN. In stars, the bottleneck is passed by triple collisions of helium-4 nuclei, producing carbon (the triple-alpha process). However, this process is very slow and requires much higher densities, taking tens of thousands of years to convert a significant amount of helium to carbon in stars, and therefore it made a negligible contribution in the minutes following the Big Bang.

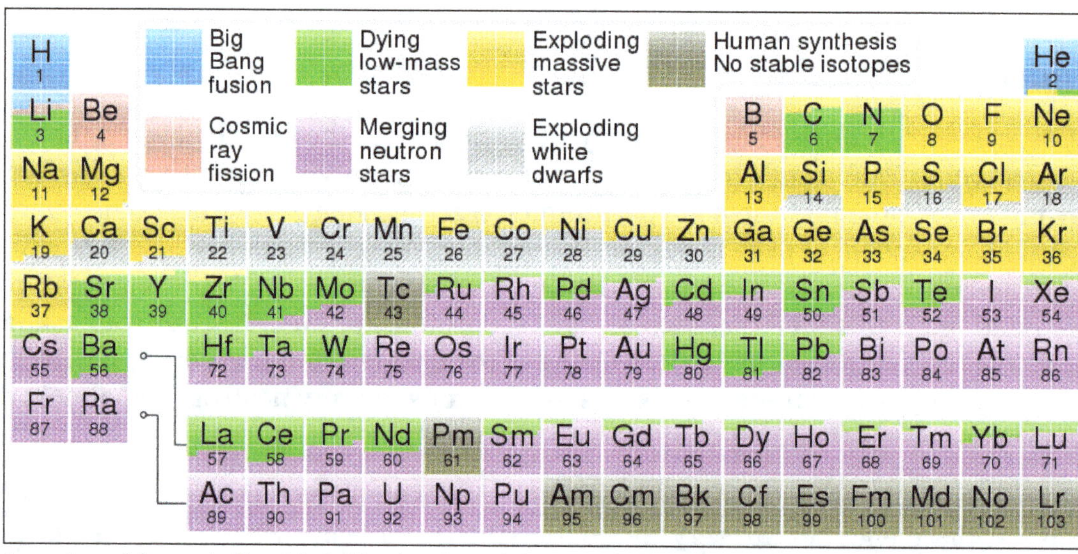

A version of the periodic table indicating the origins – including big bang nucleosynthesis – of the elements. All elements above 103 (lawrencium) are also manmade and are not included.

The predicted abundance of CNO isotopes produced in Big Bang nucleosynthesis is expected to be on the order of 10^{-15} that of H, making them essentially undetectable and negligible. Indeed, none of these primordial isotopes of the elements from beryllium to oxygen have yet been detected, although those of beryllium and boron may be able to be detected in the future. So far, the only stable nuclides known experimentally to have been made before or during Big Bang nucleosynthesis are protium, deuterium, helium-3, helium-4, and lithium-7.

Helium-4

Big Bang nucleosynthesis predicts a primordial abundance of about 25% helium-4 by mass, irrespective of the initial conditions of the universe. As long as the universe was hot enough for protons and neutrons to transform into each other easily, their ratio, determined solely by their relative masses, was about 1 neutron to 7 protons (allowing for some decay of neutrons into protons). Once it was cool enough, the neutrons quickly bound with an equal number of protons to form first deuterium, then helium-4. Helium-4 is very stable and is nearly the end of this chain if it runs for only a short time, since helium neither decays nor combines easily to form heavier nuclei (since there are no stable nuclei with mass numbers of 5 or 8, helium does not combine easily with either protons, or with itself). Once temperatures are lowered, out of every 16 nucleons (2 neutrons and 14 protons), 4 of these (25% of the total particles and total mass) combine quickly into one helium-4 nucleus. This produces one helium for every 12 hydrogens, resulting in a universe that is a little over 8% helium by number of atoms, and 25% helium by mass.

One analogy is to think of helium-4 as ash, and the amount of ash that one forms when one completely burns a piece of wood is insensitive to how one burns it. The resort to the BBN theory of the helium-4 abundance is necessary as there is far more helium-4 in the universe than can be explained by stellar nucleosynthesis. In addition, it provides an important test for the Big Bang theory. If the observed helium abundance is significantly different from 25%, then this would pose a serious challenge to the theory. This would particularly be the case if the early helium-4 abundance was much smaller than 25% because it is hard to destroy helium-4. For a few years during the mid-1990s, observations suggested that this might be the case, causing astrophysicists to talk about a Big Bang nucleosynthetic crisis, but further observations were consistent with the Big Bang theory.

Deuterium

Deuterium is in some ways the opposite of helium-4, in that while helium-4 is very stable and difficult to destroy, deuterium is only marginally stable and easy to destroy. The temperatures, time, and densities were sufficient to combine a substantial fraction of the deuterium nuclei to form helium-4 but insufficient to carry the process further using helium-4 in the next fusion step. BBN did not convert all of the deuterium in the universe to helium-4 due to the expansion that cooled the universe and reduced the density, and so cut that conversion short before it could proceed any further. One consequence of this is that, unlike helium-4, the amount of deuterium is very sensitive to initial conditions. The denser the initial universe was, the more deuterium would be converted to helium-4 before time ran out, and the less deuterium would remain.

There are no known post-Big Bang processes which can produce significant amounts of deuterium. Hence observations about deuterium abundance suggest that the universe is not infinitely old, which is in accordance with the Big Bang theory.

During the 1970s, there were major efforts to find processes that could produce deuterium, but those revealed ways of producing isotopes other than deuterium. The problem was that while the concentration of deuterium in the universe is consistent with the Big Bang model as a whole, it is too high to be consistent with a model that presumes that most of the universe is composed of protons and neutrons. If one assumes that all of the universe consists of protons and neutrons, the density of the universe is such that much of the currently observed deuterium would have been burned into helium-4. The standard explanation now used for the abundance of deuterium is that the universe does not consist mostly of baryons, but that non-baryonic matter (also known as dark matter) makes up most of the mass of the universe. This explanation is also consistent with calculations that show that a universe made mostly of protons and neutrons would be far more clumpy than is observed.

It is very hard to come up with another process that would produce deuterium other than by nuclear fusion. Such a process would require that the temperature be hot enough to produce deuterium, but not hot enough to produce helium-4, and that this process should immediately cool to non-nuclear temperatures after no more than a few minutes. It would also be necessary for the deuterium to be swept away before it reoccurs.

Producing deuterium by fission is also difficult. The problem here again is that deuterium is very unlikely due to nuclear processes, and that collisions between atomic nuclei are likely to result either in the fusion of the nuclei, or in the release of free neutrons or alpha particles. During the 1970s, cosmic ray spallation was proposed as a source of deuterium. That theory failed to account for the abundance of deuterium, but led to explanations of the source of other light elements.

Lithium

Lithium-7 and lithium-6 produced in the Big Bang are on the order of: lithium-7 to be 10^{-9} of all primordial nuclides; and lithium-6 around 10^{-13}.

Measurements and Status of Theory

The theory of BBN gives a detailed mathematical description of the production of the light "elements" deuterium, helium-3, helium-4, and lithium-7. Specifically, the theory yields precise quantitative predictions for the mixture of these elements, that is, the primordial abundances at the end of the big-bang.

In order to test these predictions, it is necessary to reconstruct the primordial abundances as faithfully as possible, for instance by observing astronomical objects in which very little stellar nucleosynthesis has taken place (such as certain dwarf galaxies) or by observing objects that are very far away, and thus can be seen in a very early stage of their evolution (such as distant quasars).

As noted above, in the standard picture of BBN, all of the light element abundances depend on the amount of ordinary matter (baryons) relative to radiation (photons). Since the universe is presumed to be homogeneous, it has one unique value of the baryon-to-photon ratio. For a long time, this meant that to test BBN theory against observations one had to ask: can all of the light element observations be explained with a single value of the baryon-to-photon ratio? Or more precisely, allowing for the finite precision of both the predictions and the observations, one asks: is there some range of baryon-to-photon values which can account for all of the observations?

More recently, the question has changed: Precision observations of the cosmic microwave background radiation with the Wilkinson Microwave Anisotropy Probe (WMAP) and Planck give an independent value for the baryon-to-photon ratio. Using this value, are the BBN predictions for the abundances of light elements in agreement with the observations?

The present measurement of helium-4 indicates good agreement, and yet better agreement for helium-3. But for lithium-7, there is a significant discrepancy between BBN and WMAP/Planck, and the abundance derived from Population II stars. The discrepancy is a factor of 2.4–4.3 below the theoretically predicted value and is considered a problem for the original models, that have resulted in revised calculations of the standard BBN based on new nuclear data, and to various reevaluation proposals for primordial proton-proton nuclear reactions, especially the abundances of $^{7}Be + n \rightarrow {}^{7}Li + p$, versus $^{7}Be + {}^{2}H \rightarrow {}^{8}Be + p$.

Non-standard Scenarios

In addition to the standard BBN scenario there are numerous non-standard BBN scenarios. These should not be confused with non-standard cosmology: a non-standard BBN scenario assumes that the Big Bang occurred, but inserts additional physics in order to see how this affects elemental abundances. These pieces of additional physics include relaxing or removing the assumption of homogeneity, or inserting new particles such as massive neutrinos.

There have been, and continue to be, various reasons for researching non-standard BBN. The first, which is largely of historical interest, is to resolve inconsistencies between BBN predictions and observations. This has proved to be of limited usefulness in that the inconsistencies were resolved by better observations, and in most cases trying to change BBN resulted in abundances that were more inconsistent with observations rather than less. The second reason for researching non-standard BBN, and largely the focus of non-standard BBN in the early 21st century, is to use BBN to place limits on unknown or speculative physics. For example, standard BBN assumes that no exotic hypothetical particles were involved in BBN. One can insert a hypothetical particle (such as a massive neutrino) and see what has to happen before BBN predicts abundances that are very different from observations. This has been done to put limits on the mass of a stable tau neutrino.

COSMIC MICROWAVE BACKGROUND

The cosmic microwave background (CMB, CMBR), in Big Bang cosmology, is electromagnetic radiation as a remnant from an early stage of the universe, also known as "relic radiation". The CMB is faint cosmic background radiation filling all space. It is an important source of data on the early universe because it is the oldest electromagnetic radiation in the universe, dating to the epoch of recombination. With a traditional optical telescope, the space between stars and galaxies (the background) is completely dark. However, a sufficiently sensitive radio telescope shows a faint background noise, or glow, almost isotropic, that is not associated with any star, galaxy, or other object. This glow is strongest in the microwave region of the radio spectrum. The accidental discovery of the CMB in 1964 by American radio astronomers Arno Penzias and Robert Wilson was the culmination of work initiated in the 1940s, and earned the discoverers the 1978 Nobel Prize in Physics.

CMB is landmark evidence of the Big Bang origin of the universe. When the universe was young, before the formation of stars and planets, it was denser, much hotter, and filled with a uniform glow from a white-hot fog of hydrogen plasma. As the universe expanded, both the plasma and the radiation filling it grew cooler. When the universe cooled enough, protons and electrons combined to form neutral hydrogen atoms. Unlike the uncombined protons and electrons, these newly conceived atoms could not scatter the thermal radiation by Thomson scattering, and so the universe became transparent instead of being an opaque fog. Cosmologists refer to the time period when neutral atoms first formed as the recombination epoch, and the event shortly afterwards when photons started to travel freely through space rather than constantly being scattered by electrons and protons in plasma is referred to as photon decoupling. The photons that existed at the time of photon decoupling have been propagating ever since, though growing fainter and less energetic, since the expansion of space causes their wavelength to increase over time (and wavelength is inversely proportional to energy according to Planck's relation). This is the source of the alternative term relic radiation. The surface of last scattering refers to the set of points in space at the right distance from us so that we are now receiving photons originally emitted from those points at the time of photon decoupling.

Tiny residual variations in the glow show a very specific pattern, as would be expected of a fairly uniformly distributed hot gas that has expanded to the current size of the universe. In particular, the spectral radiance contains small anisotropies, or irregularities, which vary with the size of the region examined. They have been measured in detail, and match what would be expected if small thermal variations, generated by quantum fluctuations of matter in a very tiny space, had expanded to the size of the observable universe we see today. Although many different processes might produce the general form of a black body spectrum, no model other than the Big Bang has yet explained the fluctuations. As a result, most cosmologists consider the Big Bang model of the universe to be the best explanation for the CMB.

Importance of Precise Measurement

Precise measurements of the CMB are critical to cosmology, since any proposed model of the universe must explain this radiation. The CMB has a thermal black body spectrum at a temperature of 2.72548±0.00057 K. The spectral radiance $dE_\nu/d\nu$ peaks at 160.23 GHz, in the microwave range of frequencies, corresponding to a photon energy of about 6.626×10^{-4} eV. Alternatively, if spectral radiance is defined as $dE_\lambda/d\lambda$, then the peak wavelength is 1.063 mm (282 GHz, 1.168×10^{-3} eV photons). The glow is very nearly uniform in all directions, but the tiny residual variations show a very specific pattern, the same as that expected of a fairly uniformly distributed hot gas that has expanded to the current size of the universe. In particular, the spectral radiance at different angles of observation in the sky contains small anisotropies, or irregularities, which vary with the size of the region examined. They have been measured in detail, and match what would be expected if small thermal variations, generated by quantum fluctuations of matter in a very tiny space, had expanded to the size of the observable universe we see today. This is a very active field of study, with scientists seeking both better data (for example, the Planck spacecraft) and better interpretations of the initial conditions of expansion. Although many different processes might produce the general form of a black body spectrum, no model other than the Big Bang has yet explained the fluctuations. As a result, most cosmologists consider the Big Bang model of the universe to be the best explanation for the CMB.

The high degree of uniformity throughout the observable universe and its faint but measured anisotropy lend strong support for the Big Bang model in general and the ΛCDM ("Lambda Cold Dark Matter") model in particular. Moreover, the fluctuations are coherent on angular scales that are larger than the apparent cosmological horizon at recombination. Either such coherence is acausally fine-tuned, or cosmic inflation occurred.

Features

The cosmic microwave background radiation is an emission of uniform, black body thermal energy coming from all parts of the sky. The radiation is isotropic to roughly one part in 100,000: the root mean square variations are only 18 μK, after subtracting out a dipole anisotropy from the Doppler shift of the background radiation. The latter is caused by the peculiar velocity of the Sun relative to the comoving cosmic rest frame as it moves at some 369.82±0.11 km/s towards the constellation Leo (galactic longitude 264.021±0.011, galactic latitude 48.253±0.005). The CMB dipole as well as aberration at higher multipoles have been measured, consistent with galactic motion.

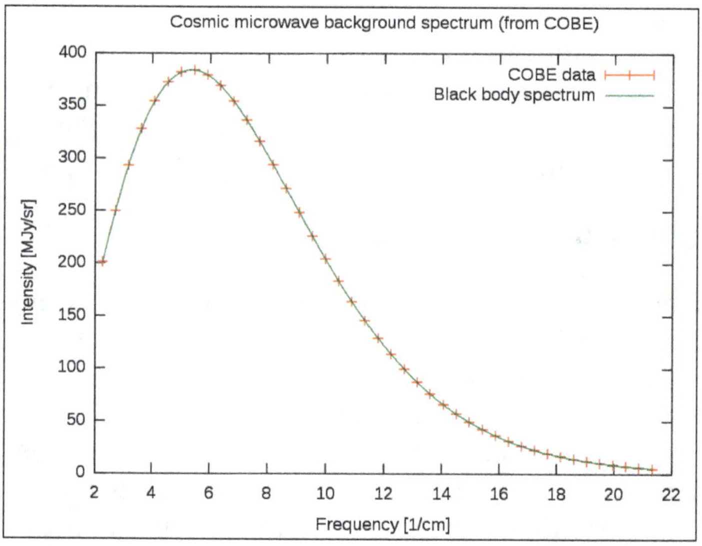

Graph of cosmic microwave background spectrum measured by the FIRAS instrument on the COBE, the most precisely measured black body spectrum in nature. The error bars are too small to be seen even in an enlarged image, and it is impossible to distinguish the observed data from the theoretical curve.

In the Big Bang model for the formation of the universe, inflationary cosmology predicts that after about 10^{-37} seconds the nascent universe underwent exponential growth that smoothed out nearly all irregularities. The remaining irregularities were caused by quantum fluctuations in the inflaton field that caused the inflation event. Long before the formation of stars and planets, the early universe was smaller, much hotter and, starting 10^{-6} seconds after the Big Bang, filled with a uniform glow from its white-hot fog of interacting plasma of photons, electrons, and baryons.

As the universe expanded, adiabatic cooling caused the energy density of the plasma to decrease until it became favorable for electrons to combine with protons, forming hydrogen atoms. This recombination event happened when the temperature was around 3000 K or when the universe was approximately 379,000 years old. As photons did not interact with these electrically neutral atoms, the former began to travel freely through space, resulting in the decoupling of matter and radiation.

The color temperature of the ensemble of decoupled photons has continued to diminish ever since; now down to 2.7260 ± 0.0013 K, it will continue to drop as the universe expands. The intensity of the radiation also corresponds to black-body radiation at 2.726 K because red-shifted black-body radiation is just like black-body radiation at a lower temperature. According to the Big Bang model, the radiation from the sky we measure today comes from a spherical surface called the surface of last scattering. This represents the set of locations in space at which the decoupling event is estimated to have occurred and at a point in time such that the photons from that distance have just reached observers. Most of the radiation energy in the universe is in the cosmic microwave background, making up a fraction of roughly 6×10^{-5} of the total density of the universe.

Two of the greatest successes of the Big Bang theory are its prediction of the almost perfect black body spectrum and its detailed prediction of the anisotropies in the cosmic microwave background. The CMB spectrum has become the most precisely measured black body spectrum in nature.

Density of energy for CMB is 0.25 eV/cm^3 (4.005×10^{-14} J/m^3) or (400–500 photons/cm^3).

Relationship to the Big Bang

The cosmic microwave background radiation and the cosmological redshift-distance relation are together regarded as the best available evidence for the Big Bang theory. Measurements of the CMB have made the inflationary Big Bang theory the Standard Cosmological Model. The discovery of the CMB in the mid-1960s curtailed interest in alternatives such as the steady state theory.

The CMB essentially confirms the Big Bang theory. In the late 1940s Alpher and Herman reasoned that if there was a big bang, the expansion of the universe would have stretched and cooled the high-energy radiation of the very early universe into the microwave region of the electromagnetic spectrum, and down to a temperature of about 5 K. They were slightly off with their estimate, but they had exactly the right idea. They predicted the CMB. It took another 15 years for Penzias and Wilson to stumble into discovering that the microwave background was actually there.

The CMB gives a snapshot of the universe when, according to standard cosmology, the temperature dropped enough to allow electrons and protons to form hydrogen atoms, thereby making the universe nearly transparent to radiation because light was no longer being scattered off free electrons. When it originated some 380,000 years after the Big Bang—this time is generally known as the "time of last scattering" or the period of recombination or decoupling—the temperature of the universe was about 3000 K. This corresponds to an energy of about 0.26 eV, which is much less than the 13.6 eV ionization energy of hydrogen.

Since decoupling, the temperature of the background radiation has dropped by a factor of roughly 1,100 due to the expansion of the universe. As the universe expands, the CMB photons are redshifted, causing them to decrease in energy. The temperature of this radiation stays inversely proportional to a parameter that describes the relative expansion of the universe over time, known as the scale length. The temperature Tr of the CMB as a function of redshift, z, can be shown to be proportional to the temperature of the CMB as observed in the present day (2.725 K or 0.2348 meV):

$$T_r = 2.725 \cdot (1 + z).$$

Primary Anisotropy

The anisotropy, or directional dependency, of the cosmic microwave background is divided into two types: primary anisotropy, due to effects that occur at the last scattering surface and before; and secondary anisotropy, due to effects such as interactions of the background radiation with hot gas or gravitational potentials, which occur between the last scattering surface and the observer.

The structure of the cosmic microwave background anisotropies is principally determined by two effects: acoustic oscillations and diffusion damping (also called collisionless damping or Silk damping). The acoustic oscillations arise because of a conflict in the photon–baryon plasma in the early universe. The pressure of the photons tends to erase anisotropies, whereas the gravitational attraction of the baryons, moving at speeds much slower than light, makes them tend to collapse to form overdensities. These two effects compete to create acoustic oscillations, which give the microwave background its characteristic peak structure. The peaks correspond, roughly, to resonances in which the photons decouple when a particular mode is at its peak amplitude.

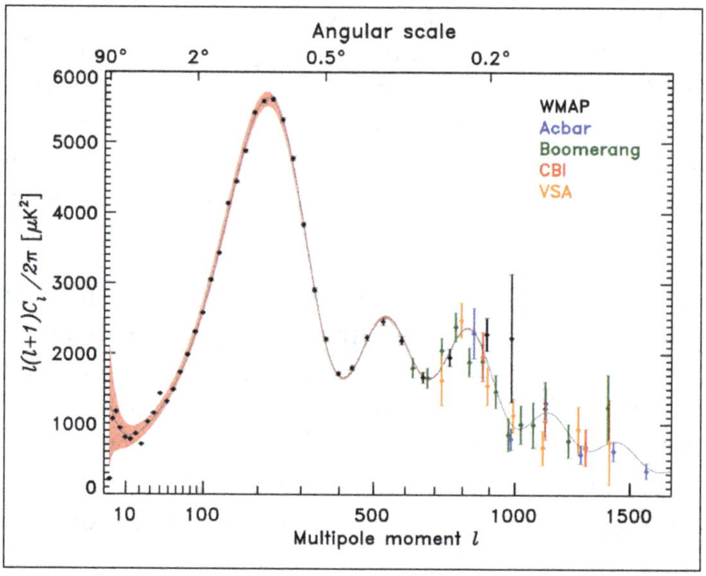

The power spectrum of the cosmic microwave background radiation temperature anisotropy in terms of the angular scale (or multipole moment). The data shown comes from the WMAP, Acbar Boomerang, CBI, and VSA instruments. Also shown is a theoretical model (solid line).

The peaks contain interesting physical signatures. The angular scale of the first peak determines the curvature of the universe (but not the topology of the universe). The next peak—ratio of the odd peaks to the even peaks—determines the reduced baryon density. The third peak can be used to get information about the dark-matter density.

The locations of the peaks also give important information about the nature of the primordial density perturbations. There are two fundamental types of density perturbations called adiabatic and isocurvature. A general density perturbation is a mixture of both, and different theories that purport to explain the primordial density perturbation spectrum predict different mixtures.

- Adiabatic density perturbations: In an adiabatic density perturbation, the fractional additional number density of each type of particle (baryons, photons, etc.) is the same. That is, if at one place there is a 1% higher number density of baryons than average, then at that place there is also a 1% higher number density of photons (and a 1% higher number density in neutrinos) than average. Cosmic inflation predicts that the primordial perturbations are adiabatic.
- Isocurvature density perturbations: In an isocurvature density perturbation, the sum (over different types of particle) of the fractional additional densities is zero. That is, a perturbation where at some spot there is 1% more energy in baryons than average, 1% more energy in photons than average, and 2% less energy in neutrinos than average, would be a pure isocurvature perturbation. Cosmic strings would produce mostly isocurvature primordial perturbations.

The CMB spectrum can distinguish between these two because these two types of perturbations produce different peak locations. Isocurvature density perturbations produce a series of peaks whose angular scales (l values of the peaks) are roughly in the ratio 1:3:5:..., while adiabatic density perturbations produce peaks whose locations are in the ratio 1:2:3:... Observations are consistent with the primordial density perturbations being entirely adiabatic, providing key support for inflation, and ruling out many models of structure formation involving, for example, cosmic strings.

Collisionless damping is caused by two effects, when the treatment of the primordial plasma as fluid begins to break down:

- The increasing mean free path of the photons as the primordial plasma becomes increasingly rarefied in an expanding universe,
- The finite depth of the last scattering surface (LSS), which causes the mean free path to increase rapidly during decoupling, even while some Compton scattering is still occurring.

These effects contribute about equally to the suppression of anisotropies at small scales and give rise to the characteristic exponential damping tail seen in the very small angular scale anisotropies.

The depth of the LSS refers to the fact that the decoupling of the photons and baryons does not happen instantaneously, but instead requires an appreciable fraction of the age of the universe up to that era. One method of quantifying how long this process took uses the photon visibility function (PVF). This function is defined so that, denoting the PVF by P(t), the probability that a CMB photon last scattered between time t and t + dt is given by P(t) dt.

The maximum of the PVF (the time when it is most likely that a given CMB photon last scattered) is known quite precisely. The first-year WMAP results put the time at which P(t) has a maximum as 372,000 years. This is often taken as the "time" at which the CMB formed. However, to figure out how long it took the photons and baryons to decouple, we need a measure of the width of the PVF. The WMAP team finds that the PVF is greater than half of its maximal value (the "full width at half maximum", or FWHM) over an interval of 115,000 years. By this measure, decoupling took place over roughly 115,000 years, and when it was complete, the universe was roughly 487,000 years old.

Late Time Anisotropy

Since the CMB came into existence, it has apparently been modified by several subsequent physical processes, which are collectively referred to as late-time anisotropy, or secondary anisotropy. When the CMB photons became free to travel unimpeded, ordinary matter in the universe was mostly in the form of neutral hydrogen and helium atoms. However, observations of galaxies today seem to indicate that most of the volume of the intergalactic medium (IGM) consists of ionized material (since there are few absorption lines due to hydrogen atoms). This implies a period of reionization during which some of the material of the universe was broken into hydrogen ions.

The CMB photons are scattered by free charges such as electrons that are not bound in atoms. In an ionized universe, such charged particles have been liberated from neutral atoms by ionizing (ultraviolet) radiation. Today these free charges are at sufficiently low density in most of the volume of the universe that they do not measurably affect the CMB. However, if the IGM was ionized at very early times when the universe was still denser, then there are two main effects on the CMB:

1. Small scale anisotropies are erased. (Just as when looking at an object through fog, details of the object appear fuzzy).

2. The physics of how photons are scattered by free electrons (Thomson scattering) induces polarization anisotropies on large angular scales. This broad angle polarization is correlated with the broad angle temperature perturbation.

Both of these effects have been observed by the WMAP spacecraft, providing evidence that the universe was ionized at very early times, at a redshift more than 17. The detailed provenance of this early ionizing radiation is still a matter of scientific debate. It may have included starlight from the very first population of stars (population III stars), supernovae when these first stars reached the end of their lives, or the ionizing radiation produced by the accretion disks of massive black holes.

The time following the emission of the cosmic microwave background—and before the observation of the first stars—is semi-humorously referred to by cosmologists as the Dark Age, and is a period which is under intense study by astronomers.

Two other effects which occurred between reionization and our observations of the cosmic microwave background, and which appear to cause anisotropies, are the Sunyaev–Zel'dovich effect, where a cloud of high-energy electrons scatters the radiation, transferring some of its energy to the CMB photons, and the Sachs–Wolfe effect, which causes photons from the Cosmic Microwave Background to be gravitationally redshifted or blueshifted due to changing gravitational fields.

Polarization

The cosmic microwave background is polarized at the level of a few microkelvin. There are two types of polarization, called E-modes and B-modes. This is in analogy to electrostatics, in which the electric field (E-field) has a vanishing curl and the magnetic field (B-field) has a vanishing divergence. The E-modes arise naturally from Thomson scattering in a heterogeneous plasma. The B-modes are not produced by standard scalar type perturbations. Instead they can be created by two mechanisms: the first one is by gravitational lensing of E-modes, which has been measured by the South Pole Telescope in 2013; the second one is from gravitational waves arising from cosmic inflation. Detecting the B-modes is extremely difficult, particularly as the degree of foreground contamination is unknown, and the weak gravitational lensing signal mixes the relatively strong E-mode signal with the B-mode signal.

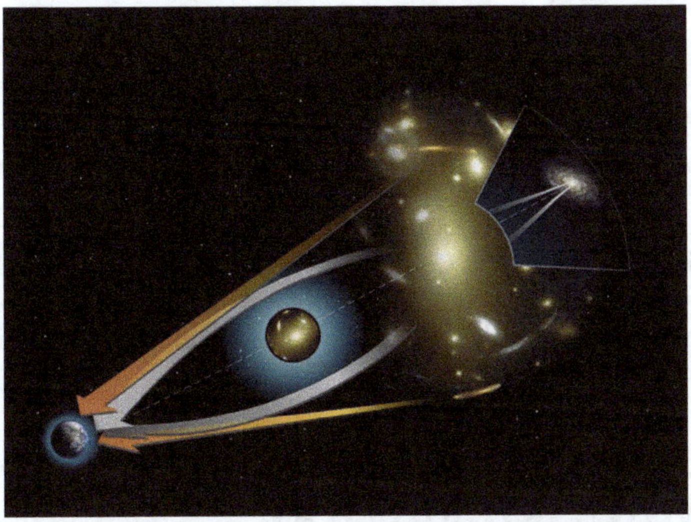

This artist's impression shows how light from the early universe is deflected by the gravitational lensing effect of massive cosmic structures forming B-modes as it travels across the universe.

- E-modes: E-modes were first seen in 2002 by the Degree Angular Scale Interferometer (DASI).
- B-modes: Cosmologists predict two types of B-modes, the first generated during cosmic inflation shortly after the big bang, and the second generated by gravitational lensing at later times.

Primordial Gravitational Waves

Primordial gravitational waves are gravitational waves that could be observed in the polarisation of the cosmic microwave background and having their origin in the early universe. Models of cosmic inflation predict that such gravitational waves should appear; thus, their detection supports the theory of inflation, and their strength can confirm and exclude different models of inflation. It is the result of three things: inflationary expansion of space itself, reheating after inflation, and turbulent fluid mixing of matter and radiation.

On 17 March 2014 it was announced that the BICEP2 instrument had detected the first type of B-modes, consistent with inflation and gravitational waves in the early universe at the level of $r = 0.20\,^{+0.07}_{-0.05}$, which is the amount of power present in gravitational waves compared to the amount of power present in other scalar density perturbations in the very early universe. Had this been confirmed it would have provided strong evidence of cosmic inflation and the Big Bang, but on 19 June 2014, considerably lowered confidence in confirming the findings was reported and on 19 September 2014 new results of the Planck experiment reported that the results of BICEP2 can be fully attributed to cosmic dust.

Gravitational Lensing

The second type of B-modes was discovered in 2013 using the South Pole Telescope with help from the Herschel Space Observatory. This discovery may help test theories on the origin of the universe. Scientists are using data from the Planck mission by the European Space Agency, to gain a better understanding of these waves.

In October 2014, a measurement of the B-mode polarization at 150 GHz was published by the POLARBEAR experiment. Compared to BICEP2, POLARBEAR focuses on a smaller patch of the sky and is less susceptible to dust effects. The team reported that POLARBEAR's measured B-mode polarization was of cosmological origin (and not just due to dust) at a 97.2% confidence level.

STRUCTURE FORMATION

In physical cosmology, structure formation is the formation of galaxies, galaxy clusters and larger structures from small early density fluctuations. The universe, as is now known from observations of the cosmic microwave background radiation, began in a hot, dense, nearly uniform state approximately 13.8 billion years ago. However, looking in the sky today, we see structures on all scales, from stars and planets to galaxies and, on still larger scales, galaxy clusters and sheet-like structures of galaxies separated by enormous voids containing few galaxies. Structure formation attempts to model how these structures formed by gravitational instability of small early density ripples.

The modern Lambda-CDM model is successful at predicting the observed large-scale distribution of galaxies, clusters and voids; but on the scale of individual galaxies there are many complications due to highly nonlinear processes involving baryonic physics, gas heating and cooling, star formation and feedback. Understanding the processes of galaxy formation is a major topic of modern cosmology research, both via observations such as the Hubble Ultra-Deep Field and via large computer simulations.

Under present models, the structure of the visible universe was formed in the following stages:

Very Early Universe

In this stage, some mechanism, such as cosmic inflation, was responsible for establishing the initial conditions of the universe: homogeneity, isotropy, and flatness. Cosmic inflation also would have amplified minute quantum fluctuations (pre-inflation) into slight density ripples of overdensity and underdensity (post-inflation).

Growth of Structure

The early universe was dominated by radiation; in this case density fluctuations larger than the cosmic horizon grow proportional to the scale factor, as the gravitational potential fluctuations remain constant. Structures smaller than the horizon remained essentially frozen due to radiation domination impeding growth. As the universe expanded, the density of radiation drops faster than matter (due to redshifting of photon energy); this led to a crossover called matter-radiation equality at ~ 50,000 years after the Big Bang. After this all dark matter ripples could grow freely, forming seeds into which the baryons could later fall. The size of the universe at this epoch forms a turnover in the matter power spectrum which can be measured in large redshift surveys.

Recombination

The universe was dominated by radiation for most of this stage, and due to the intense heat and radiation, the primordial hydrogen and helium were fully ionized into nuclei and free electrons. In this hot and dense situation, the radiation (photons) could not travel far before Thomson scattering off an electron. The universe was very hot and dense, but expanding rapidly and therefore cooling. Finally, at a little less than 400,000 years after the 'bang', it became cool enough (around 3000 K) for the protons to capture negatively charged electrons, forming neutral hydrogen atoms. (Helium atoms formed somewhat earlier due to their larger binding energy). Once nearly all the charged particles were bound in neutral atoms, the photons no longer interacted with them and were free to propagate for the next 13.8 billion years; we currently detect those photons redshifted by a factor 1090 down to 2.725 K as the Cosmic Microwave Background Radiation (CMB) filling today's universe. Several remarkable space-based missions (COBE, WMAP, Planck), have detected very slight variations in the density and temperature of the CMB. These variations were subtle, and the CMB appears very nearly uniformly the same in every direction. However, the slight temperature variations of order a few parts in 100,000 are of enormous importance, for they essentially were early "seeds" from which all subsequent complex structures in the universe ultimately developed.

The theory of what happened after the universe's first 400,000 years is one of hierarchical structure formation: the smaller gravitationally bound structures such as matter peaks containing the first stars and stellar clusters formed first, and these subsequently merged with gas and dark matter to form galaxies, followed by groups, clusters and superclusters of galaxies.

The very early universe is still a poorly understood epoch, from the viewpoint of fundamental physics. The prevailing theory, cosmic inflation, does a good job explaining the observed flatness, homogeneity and isotropy of the universe, as well as the absence of exotic relic particles (such as magnetic monopoles). Another prediction borne out by observation is that tiny perturbations in the primordial universe seed the later formation of structure. These fluctuations, while they form the foundation for all structure, appear most clearly as tiny temperature fluctuations at one part in 100,000. (To put this in perspective, the same level of fluctuations on a topographic map of the United States would show no feature taller than a few centimeters.) These fluctuations are critical, because they provide the seeds from which the largest structures can grow and eventually collapse to form galaxies and stars. COBE (Cosmic Background Explorer) provided the first detection of the intrinsic fluctuations in the cosmic microwave background radiation in the 1990s.

These perturbations are thought to have a very specific character: they form a Gaussian random field whose covariance function is diagonal and nearly scale-invariant. Observed fluctuations appear to have exactly this form, and in addition the spectral index measured by WMAP—the spectral index measures the deviation from a scale-invariant spectrum—is very nearly the value predicted by the simplest and most robust models of inflation. Another important property of the primordial perturbations, that they are adiabatic (or isentropic between the various kinds of matter that compose the universe), is predicted by cosmic inflation and has been confirmed by observations.

Other theories of the very early universe have been proposed that are claimed to make similar predictions, such as the brane gas cosmology, cyclic model, pre-big bang model and holographic universe, but they remain nascent and are not widely accepted. Some theories, such as cosmic strings, have largely been refuted by increasingly precise data.

The Horizon Problem

The physical wavelength of a perturbation mode (dashed line) is shown as well. The plot illustrates how the perturbation mode exits the horizon during cosmic inflation in order to reenter during radiation domination. If cosmic inflation never happened, and radiation domination continued back until a gravitational singularity, then the mode would never have exited the horizon in the very early universe.

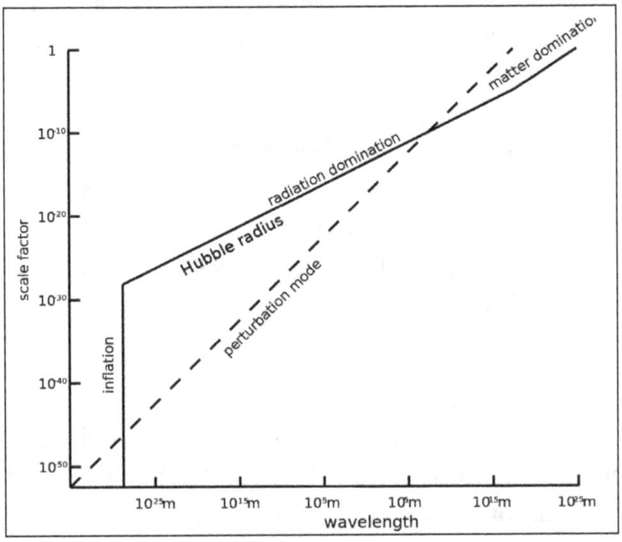

The physical size of the Hubble radius (solid line) as a function of the scale factor of the universe.

An important concept in structure formation is the notion of the Hubble radius, often called simply the horizon, as it is closely related to the particle horizon. The Hubble radius, which is related to the Hubble parameter H as $R = c / H$, where c is the speed of light, defines, roughly speaking, the volume of the nearby universe that has recently (in the last expansion time) been in causal contact with an observer. Since the universe is continually expanding, its energy density is continually decreasing (in the absence of truly exotic matter such as phantom energy). The Friedmann equation relates the energy density of the universe to the Hubble parameter and shows that the Hubble radius is continually increasing.

The horizon problem of big bang cosmology says that, without inflation, perturbations were never in causal contact before they entered the horizon and thus the homogeneity and isotropy of, for example, the large scale galaxy distributions cannot be explained. This is because, in an ordinary Friedmann–Lemaître–Robertson–Walker cosmology, the Hubble radius increases more rapidly than space expands, so perturbations only enter the Hubble radius, and are not pushed out by the expansion. This paradox is resolved by cosmic inflation, which suggests that during a phase of rapid expansion in the early universe the Hubble radius was nearly constant. Thus, large scale isotropy is due to quantum fluctuations produced during cosmic inflation that are pushed outside the horizon.

Primordial Plasma

The end of inflation is called reheating, when the inflation particles decay into a hot, thermal plasma of other particles. In this epoch, the energy content of the universe is entirely radiation, with standard model particles having relativistic velocities. As the plasma cools, baryogenesis and leptogenesis are thought to occur, as the quark–gluon plasma cools, electroweak symmetry breaking occurs and the universe becomes principally composed of ordinary protons, neutrons and electrons. As the universe cools further, Big Bang nucleosynthesis occurs and small quantities of deuterium, helium and lithium nuclei are created. As the universe cools and expands, the energy in photons begins to redshift away, particles become non-relativistic and ordinary matter begins to dominate the universe. Eventually, atoms begin to form as free electrons bind to nuclei. This suppresses Thomson scattering of photons. Combined with the rarefaction of the universe (and consequent increase in the mean free path of photons), this makes the universe transparent and the cosmic microwave background is emitted at recombination (the surface of last scattering).

Acoustic Oscillations

The primordial plasma would have had very slight overdensities of matter, thought to have derived from the enlargement of quantum fluctuations during inflation. Whatever the source, these overdensities gravitationally attract matter. But the intense heat of the near constant photon-matter interactions of this epoch rather forcefully seeks thermal equilibrium, which creates a large amount of outward pressure. These counteracting forces of gravity and pressure create oscillations, analogous to sound waves created in air by pressure differences.

These perturbations are important, as they are responsible for the subtle physics that result in the cosmic microwave background anisotropy. In this epoch, the amplitude of perturbations that enter the horizon oscillate sinusoidally, with dense regions becoming more rarefied and then becoming dense again, with a frequency which is related to the size of the perturbation. If the perturbation oscillates an integral or half-integral number of times between coming into the horizon and recombination, it appears as an acoustic peak of the cosmic microwave background anisotropy. (A half-oscillation, in which a dense region becomes a rarefied region or vice versa, appears as a peak because the anisotropy is displayed as a power spectrum, so underdensities contribute to the power just as much as overdensities.) The physics that determines the detailed peak structure of the microwave background is complicated, but these oscillations provide the essence.

Linear Structure

One of the key realizations made by cosmologists in the 1970s and 1980s was that the majority of the matter content of the universe was composed not of atoms, but rather a mysterious form of matter known as dark matter. Dark matter interacts through the force of gravity, but it is not composed of baryons, and it is known with very high accuracy that it does not emit or absorb radiation. It may be composed of particles that interact through the weak interaction, such as neutrinos, but it cannot be composed entirely of the three known kinds of neutrinos (although some have suggested it is a sterile neutrino). Recent evidence indicates that there are about five times as much dark matter as baryonic matter, and thus the dynamics of the universe in this epoch are dominated by dark matter.

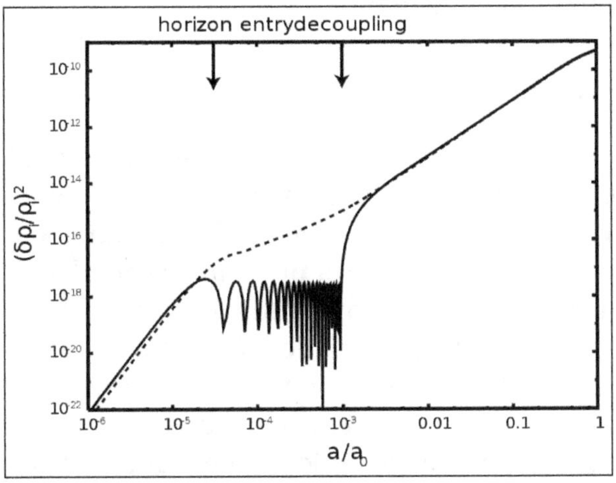

The image above shows the evolution of two perturbations to the ΛCDM homogeneous big bang model. Between entering the horizon and decoupling, the dark matter perturbation (dashed line) grows logarithmically, before the growth accelerates in matter domination. On the other hand, between entering the horizon and decoupling, the perturbation in the baryon-photon fluid (solid line) oscillates rapidly. After decoupling, it grows rapidly to match the dominant matter perturbation, the dark matter mode.

Dark matter plays a crucial role in structure formation because it feels only the force of gravity: the gravitational Jeans instability which allows compact structures to form is not opposed by any force, such as radiation pressure. As a result, dark matter begins to collapse into a complex network of dark matter halos well before ordinary matter, which is impeded by pressure forces. Without dark matter, the epoch of galaxy formation would occur substantially later in the universe than is observed.

The physics of structure formation in this epoch is particularly simple, as dark matter perturbations with different wavelengths evolve independently. As the Hubble radius grows in the expanding universe, it encompasses larger and larger disturbances. During matter domination, all causal dark matter perturbations grow through gravitational clustering. However, the shorter-wavelength perturbations that are included during radiation domination have their growth retarded until matter domination. At this stage, luminous, baryonic matter is expected to mirror the evolution of the dark matter simply, and their distributions should closely trace one another.

It is a simple matter to calculate this "linear power spectrum" and, as a tool for cosmology, it is of comparable importance to the cosmic microwave background. Galaxy surveys have measured the power spectrum, such as the Sloan Digital Sky Survey, and by surveys of the Lyman-α forest. Since these studies observe radiation emitted from galaxies and quasars, they do not directly measure the dark matter, but the large-scale distribution of galaxies (and of absorption lines in the Lyman-α forest) is expected to mirror the distribution of dark matter closely. This depends on the fact that galaxies will be larger and more numerous in denser parts of the universe, whereas they will be comparatively scarce in rarefied regions.

Nonlinear Structure

When the perturbations have grown sufficiently, a small region might become substantially denser than the mean density of the universe. At this point, the physics involved becomes substantially more complicated. When the deviations from homogeneity are small, the dark matter may be treated as a pressureless fluid and evolves by very simple equations. In regions which are significantly denser than the background, the full Newtonian theory of gravity must be included. (The Newtonian theory is appropriate because the masses involved are much less than those required to form a black hole, and the speed of gravity may be ignored as the light-crossing time for the structure is still smaller than the characteristic dynamical time.) One sign that the linear and fluid approximations become invalid is that dark matter starts to form caustics in which the trajectories of adjacent particles cross, or particles start to form orbits. These dynamics are best understood using N-body simulations (although a variety of semi-analytic schemes, such as the Press–Schechter formalism, can be used in some cases). While in principle these simulations are quite simple, in practice they are tough to implement, as they require simulating millions or even billions of particles. Moreover, despite the large number of particles, each particle typically weighs 109 solar masses and discretization effects may become significant. The largest such simulation as of 2005 is the Millennium simulation.

The result of N-body simulations suggests that the universe is composed largely of voids, whose densities might be as low as one-tenth the cosmological mean. The matter condenses in large filaments and haloes which have an intricate web-like structure. These form galaxy groups, clusters and superclusters. While the simulations appear to agree broadly with observations, their interpretation is complicated by the understanding of how dense accumulations of dark matter spur galaxy formation. In particular, many more small haloes form than we see in astronomical observations as dwarf galaxies and globular clusters. This is known as the galaxy bias problem, and a variety of explanations have been proposed. Most account for it as an effect in the complicated physics of galaxy formation, but some have suggested that it is a problem with our model of dark matter and that some effect, such as warm dark matter, prevents the formation of the smallest haloes.

Gas Evolution

The final stage in evolution comes when baryons condense in the centres of galaxy haloes to form galaxies, stars and quasars. Dark matter greatly accelerates the formation of dense haloes. As dark matter does not have radiation pressure, the formation of smaller structures from dark matter is impossible. This is because dark matter cannot dissipate angular momentum, whereas ordinary baryonic matter can collapse to form dense objects by dissipating angular momentum through radiative cooling. Understanding these processes is an enormously difficult computational problem, because they can involve the physics of gravity, magnetohydrodynamics, atomic physics, nuclear reactions, turbulence and even general relativity. In most cases, it is not yet possible to perform simulations that can be compared quantitatively with observations, and the best that can be achieved are approximate simulations that illustrate the main qualitative features of a process such as a star formation.

Modelling Structure Formation

Cosmological Perturbations

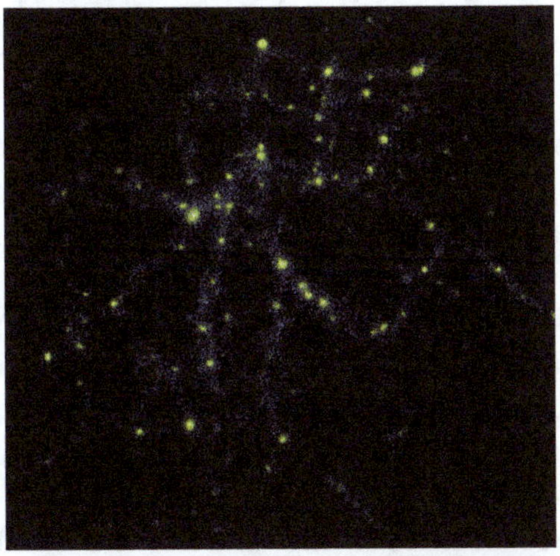

Snapshot from a computer simulation of large scale structure formation in a Lambda-CDM universe.

Much of the difficulty, and many of the disputes, in understanding the large-scale structure of the universe can be resolved by better understanding the choice of gauge in general relativity. By the scalar-vector-tensor decomposition, the metric includes four scalar perturbations, two vector perturbations, and one tensor perturbation. Only the scalar perturbations are significant: the vectors are exponentially suppressed in the early universe, and the tensor mode makes only a small (but important) contribution in the form of primordial gravitational radiation and the B-modes of the cosmic microwave background polarization. Two of the four scalar modes may be removed by a physically meaningless coordinate transformation. Which modes are eliminated determine the infinite number of possible gauge fixings. The most popular gauge is

Newtonian gauge (and the closely related conformal Newtonian gauge), in which the retained scalars are the Newtonian potentials Φ and Ψ, which correspond exactly to the Newtonian potential energy from Newtonian gravity. Many other gauges are used, including synchronous gauge, which can be an efficient gauge for numerical computation (it is used by CMBFAST). Each gauge still includes some unphysical degrees of freedom. There is a so-called gauge-invariant formalism, in which only gauge invariant combinations of variables are considered.

Inflation and Initial Conditions

The initial conditions for the universe are thought to arise from the scale invariant quantum mechanical fluctuations of cosmic inflation. The perturbation of the background energy density at a given point $\rho(\mathrm{x},t)$ in space is then given by an isotropic, homogeneous Gaussian random field of mean zero. This means that the spatial Fourier transform of $\rho - \hat{\rho}(\mathrm{k},t)$ has the following correlation functions

$$\langle \hat{\rho}\,(\mathrm{k},t)\,\hat{\rho}\,(\mathrm{k}',t)\rangle = f(k)\,\delta^{(3)}\,(\mathrm{k}-\mathrm{k}'),$$

where $\delta^{(3)}$ is the three-dimensional Dirac delta function and $k = |\,\mathrm{k}\,|$ is the length of k. Moreover, the spectrum predicted by inflation is nearly scale invariant, which means:

$$\langle \hat{\rho}\,(\mathbf{k},t)\,\hat{\rho}\,(\mathbf{k}',t)\rangle = k^{n_s-1}\delta^{(3)}\,(\mathbf{k}-\mathbf{k}'),$$

where $n_s - 1$ is a small number. Finally, the initial conditions are adiabatic or isentropic, which means that the fractional perturbation in the entropy of each species of particle is equal. The resulting predictions fit very well with observations, however there is a conceptual problem with the physical picture presented above. The quantum state from which the quantum fluctuations are extracted, is in fact completely homogeneous and isotropic, and thus it can not be argued that the quantum fluctuations represent the primordial inhomogeneities and anisotropies. The interpretation of quantum uncertainties in the value of the inflation field (which is what the so-called quantum fluctuations really are) as if they were statistical fluctuations in a Gaussian random field does not follow from the application of standard rules of quantum theory. The issue is sometimes presented in terms of the "quantum to classical transition", which is a confusing manner to refer to the problem at hand, as there are very few physicists, if any, that would argue that there is any entity that is truly classical at the fundamental level. In fact, the consideration of these issues brings us face to face with the so called measurement problem in quantum theory. If anything, the problem becomes exacerbated in the cosmological context, as the early universe contains no entities that might be taken as playing the role of "observers" or of "measuring devices", both of which are essential for the standard usage of quantum mechanics. The most popular posture among cosmologists, in this regard, is to rely on arguments based on decoherence and some form of "Many Worlds

Interpretation" of quantum theory. There is an intense ongoing debate about the reasonableness of that posture.

Galaxy Filament

In physical cosmology, galaxy filaments (subtypes: supercluster complexes, galaxy walls, and galaxy sheets) are the largest known structures in the universe. They are massive, thread-like formations, with a typical length of 50 to 80 megaparsecs h^{-1} (or of the order of 200 to 500 million light-years) that form the boundaries between large voids in the universe. Filaments consist of gravitationally bound galaxies. Parts wherein many galaxies are very close to one another (in cosmic terms) are called superclusters.

Formation

In the standard model of the evolution of the universe, galactic filaments form along and follow web-like strings of dark matter. It is thought that this dark matter dictates the structure of the Universe on the grandest of scales. Dark matter gravitationally attracts baryonic matter, and it is this "normal" matter that astronomers see forming long, thin walls of super-galactic clusters.

Discovery

Discovery of structures larger than superclusters began in the late-1980s. In 1987, astronomer R. Brent Tully of the University of Hawaii's Institute of Astronomy identified what he called the Pisces–Cetus Supercluster Complex. In 1989, the CfA2 Great Wall was discovered, followed by the Sloan Great Wall in 2003. On January 11, 2013, researchers led by Roger Clowes of the University of Central Lancashire announced the discovery of a large quasar group, the Huge-LQG, which dwarfs previously discovered galaxy filaments in size. In November 2013, using gamma-ray bursts as reference points, astronomers discovered the Hercules–Corona Borealis Great Wall, an extremely huge filament measuring more than 10 billion light-years across.

Filaments

Filament subtype of filaments have roughly similar major and minor axes in cross-section, along the lengthwise axis.

Filaments of Galaxies			
Filament	Date	Mean distance	Dimension
Coma Filament			
Perseus–Pegasus Filament	1985		
Ursa Major Filament			

Lynx–Ursa Major Filament (LUM Filament)	1999	from 2000 km/s to 8000 km/s in redshift space	
z=2.38 filament around protocluster ClG J2143-4423	2004	z=2.38	110Mpc

A short filament, detected by identifying an alignment of star-forming galaxies, in the neighborhood of the Milky Way and the Local Group was proposed by Adi Zitrin and Noah Brosch. The reality of this filament, and the identification of a similar but shorter filament, were the result of a study by McQuinn *et al.* based on distance measurements using the TRGB method.

Galaxy Walls

The galaxy wall subtype of filaments have a significantly greater major axis than minor axis in cross-section, along the lengthwise axis.

Walls of Galaxies			
Wall	Date	Mean distance	Dimension
CfA2 Great Wall (Coma Wall, Great Wall, Northern Great Wall, Great Northern Wall, CfA Great Wall)	1989	z=0.03058	251Mpc long 750 Mly long 250 Mly wide 20 Mly thick
Sloan Great Wall (SDSS Great Wall)	2003	z=0.07804	433Mpc long
Sculptor Wall (Southern Great Wall, Great Southern Wall, Southern Wall)			8000 km/s long 5000 km/s wide 1000 km/s deep (in redshift space dimensions)
Grus Wall			
Fornax Wall			
Hercules–Corona Borealis Great Wall	2013	z≈2	3 Gpc long, 150 000 km/s deep (in redshift space)

- A "Centaurus Great Wall" (or "Fornax Great Wall" or "Virgo Great Wall") has been proposed, which would include the Fornax Wall as a portion of it (visually created by the Zone of Avoidance) along with the Centaurus Supercluster and the Virgo Supercluster also known as our Local Supercluster within which the Milky Way galaxy is located (implying this to be the Local Great Wall).

- A wall was proposed to be the physical embodiment of the Great Attractor, with the Norma Cluster as part of it. It is sometimes referred to as the Great Attractor Wall or Norma Wall. This suggestion was superseded by the proposal of a supercluster, Laniakea, that would encompass the Great Attractor, Virgo Supercluster, Hydra-Centaurus Superclusters.

- A wall was proposed in 2000 to lie at z=1.47 in the vicinity of radio galaxy B3 0003+387.
- A wall was proposed in 2000 to lie at z=0.559 in the northern Hubble Deep Field (HDF North).

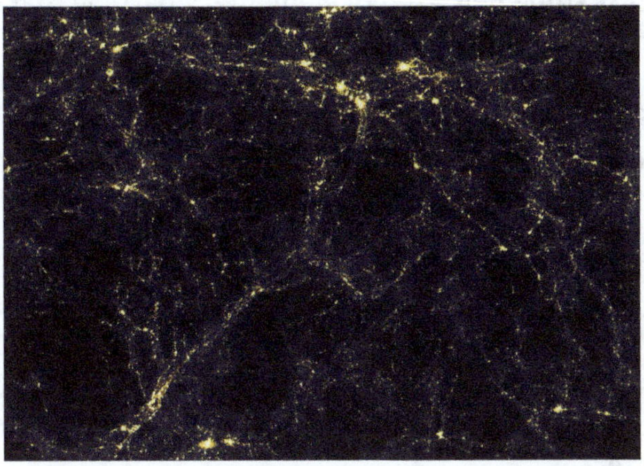

Galaxy filaments, walls and voids form web-like structures.

DARK MATTER

Dark matter is a form of matter thought to account for approximately 85% of the matter in the universe and about a quarter of its total energy density. The majority of dark matter is thought to be non-baryonic in nature, possibly being composed of some as-yet undiscovered subatomic particles. Its presence is implied in a variety of astrophysical observations, including gravitational effects which cannot be explained by accepted theories of gravity unless more matter is present than can be seen. For this reason, most experts think dark matter to be abundant in the universe and to have had a strong influence on its structure and evolution. Dark matter is called dark because it does not appear to interact with observable electromagnetic radiation, such as light, and is thus invisible to the entire electromagnetic spectrum, making it extremely difficult to detect using usual astronomical equipment.

Primary evidence for dark matter comes from calculations showing many galaxies would fly apart instead of rotating, or would not have formed or move as they do, if they did not contain a large amount of unseen matter. Other lines of evidence include observations in gravitational lensing, from the cosmic microwave background, also astronomical observations of the observable universe's current structure, the formation and evolution of galaxies, mass location during galactic collisions, and the motion of galaxies within galaxy clusters. In the standard Lambda-CDM model of cosmology, the total mass–energy of the universe contains 5% ordinary matter and energy, 27% dark matter and 68% of an unknown form of energy known as dark energy. Thus, dark matter constitutes 85% of total mass, while dark energy plus dark matter constitute 95% of total mass–energy content.

Because dark matter has not yet been observed directly, if it exists, it must barely interact with ordinary baryonic matter and radiation, except through gravity. The primary candidate for dark matter is some new kind of elementary particle that has not yet been discovered, in particular, weakly-interacting massive particles (WIMPs). Many experiments to directly detect and study dark matter particles are being actively undertaken, but none have yet succeeded. Dark matter is classified as "cold", "warm", or "hot" according to its velocity (more precisely, its free streaming length). Current models favor a cold dark matter scenario, in which structures emerge by gradual accumulation of particles.

Although the existence of dark matter is generally accepted by the scientific community, some astrophysicists, intrigued by certain observations which do not fit the dark matter theory, argue for various modifications of the standard laws of general relativity, such as modified Newtonian dynamics, tensor–vector–scalar gravity, or entropic gravity. These models attempt to account for all observations without invoking supplemental non-baryonic matter.

In standard cosmology, matter is anything whose energy density scales with the inverse cube of the scale factor, i.e., $\rho \propto a^{-3}$. This is in contrast to radiation, which scales as the inverse fourth power of the scale factor $\rho \propto a^{-4}$, and a cosmological constant, which is independent of a. These scalings can be understood intuitively: for an ordinary particle in a cubical box, doubling the length of the sides of the box decreases the density (and hence energy density) by a factor of eight (= 2^3). For radiation, the decrease in energy density is larger because an increase in scale factor causes a proportional redshift. A cosmological constant, as an intrinsic property of space, has a constant energy density regardless of the volume under consideration.

In principle, "dark matter" means all components of the universe which are not visible but still obey $\rho \propto a^{-3}$. In practice, the term "dark matter" is often used to mean only the non-baryonic component of dark matter, i.e., excluding "missing baryons." Context will usually indicate which meaning is intended.

Observational Evidence

Galaxy Rotation Curves

The arms of spiral galaxies rotate around the galactic center. The luminous mass density of a spiral galaxy decreases as one goes from the center to the outskirts. If luminous mass were all the matter, then we can model the galaxy as a point mass in the centre and test masses orbiting around it, similar to the Solar System. From Kepler's Second Law, it is expected that the rotation velocities will decrease with distance from the center, similar to the Solar System. This is not observed. Instead, the galaxy rotation curve remains flat as distance from the center increases.

If Kepler's laws are correct, then the obvious way to resolve this discrepancy is to conclude the mass distribution in spiral galaxies is not similar to that of the Solar System. In particular, there is a lot of non-luminous matter (dark matter) in the outskirts of the galaxy.

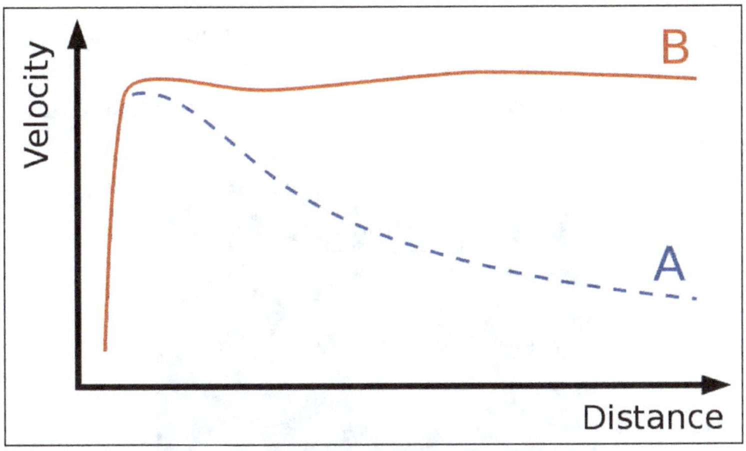

Rotation curve of a typical spiral galaxy: predicted (A) and observed (B). Dark matter can explain the 'flat' appearance of the velocity curve out to a large radius.

Velocity Dispersions

Stars in bound systems must obey the virial theorem. The theorem, together with the measured velocity distribution, can be used to measure the mass distribution in a bound system, such as elliptical galaxies or globular clusters. With some exceptions, velocity dispersion estimates of elliptical galaxies do not match the predicted velocity dispersion from the observed mass distribution, even assuming complicated distributions of stellar orbits.

As with galaxy rotation curves, the obvious way to resolve the discrepancy is to postulate the existence of non-luminous matter.

Galaxy Clusters

Galaxy clusters are particularly important for dark matter studies since their masses can be estimated in three independent ways:

- From the scatter in radial velocities of the galaxies within clusters.
- From X-rays emitted by hot gas in the clusters. From the X-ray energy spectrum and flux, the gas temperature and density can be estimated, hence giving the pressure; assuming pressure and gravity balance determines the cluster's mass profile.
- Gravitational lensing (usually of more distant galaxies) can measure cluster masses without relying on observations of dynamics (e.g., velocity).

Generally, these three methods are in reasonable agreement dark matter outweighs visible matter by approximately 5 to 1.

Gravitational Lensing

One of the consequences of general relativity is massive objects (such as a cluster of galaxies) lying between a more distant source (such as a quasar) and an observer should act as a lens to bend the light from this source. The more massive an object, the more lensing is observed.

Strong gravitational lensing as observed by the Hubble Space Telescope in Abell 1689 indicates the presence of dark matter—enlarge the image to see the lensing arcs.

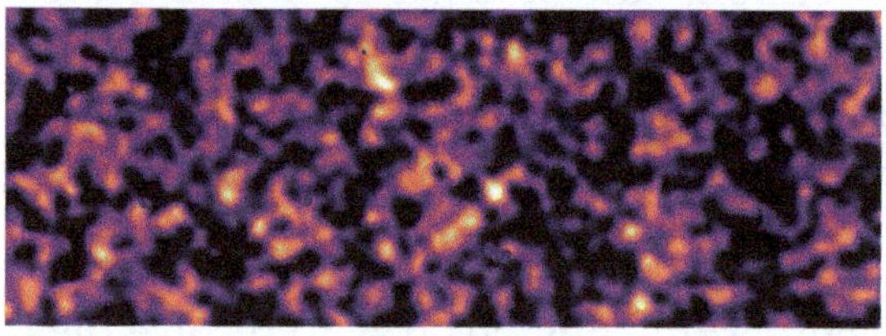

Dark matter map for a patch of sky based on gravitational lensing analysis of a Kilo-Degree survey.

Strong lensing is the observed distortion of background galaxies into arcs when their light passes through such a gravitational lens. It has been observed around many distant clusters including Abell 1689. By measuring the distortion geometry, the mass of the intervening cluster can be obtained. In the dozens of cases where this has been done, the mass-to-light ratios obtained correspond to the dynamical dark matter measurements of clusters. Lensing can lead to multiple copies of an image. By analyzing the distribution of multiple image copies, scientists have been able to deduce and map the distribution of dark matter around the MACS J0416.1-2403 galaxy cluster.

Weak gravitational lensing investigates minute distortions of galaxies, using statistical analyses from vast galaxy surveys. By examining the apparent shear deformation of the adjacent background galaxies, the mean distribution of dark matter can be characterized. The mass-to-light ratios correspond to dark matter densities predicted by other large-scale structure measurements. Dark matter does not bend light itself; mass (in this case the mass of the dark matter) bends spacetime. Light follows the curvature of spacetime, resulting in the lensing effect.

Cosmic Microwave Background

Although both dark matter and ordinary matter are matter, they do not behave in the same way. In particular, in the early universe, ordinary matter was ionized and interacted strongly with radiation via Thomson scattering. Dark matter does not interact directly with radiation, but it does affect the CMB by its gravitational potential (mainly on large scales), and by its effects on the density and velocity of ordinary matter. Ordinary and dark matter perturbations, therefore, evolve differently with time and leave different imprints on the cosmic microwave background (CMB).

The cosmic microwave background is very close to a perfect blackbody but contains very small temperature anisotropies of a few parts in 100,000. A sky map of anisotropies can be decomposed into an angular power spectrum, which is observed to contain a series of acoustic peaks at near-equal spacing but different heights. The series of peaks can be predicted for any assumed set of cosmological parameters by modern computer codes such as CMBFast and CAMB, and matching theory to data, therefore, constrains cosmological parameters. The first peak mostly shows the density of baryonic matter, while the third peak relates mostly to the density of dark matter, measuring the density of matter and the density of atoms.

The CMB anisotropy was first discovered by COBE in 1992, though this had too coarse resolution to detect the acoustic peaks. After the discovery of the first acoustic peak by the balloon-borne BOOMERanG experiment in 2000, the power spectrum was precisely observed by WMAP in 2003–2012, and even more precisely by the Planck spacecraft in 2013–2015. The results support the Lambda-CDM model.

The observed CMB angular power spectrum provides powerful evidence in support of dark matter, as its precise structure is well fitted by the Lambda-CDM model, but difficult to reproduce with any competing model such as modified Newtonian dynamics (MOND).

Structure Formation

Structure formation refers to the period after the Big Bang when density perturbations collapsed to form stars, galaxies, and clusters. Prior to structure formation, the Friedmann solutions to general relativity describe a homogeneous universe. Later, small anisotropies gradually grew and condensed the homogeneous universe into stars, galaxies and larger structures. Ordinary matter is affected by radiation, which is the dominant element of the universe at very early times. As a result, its density perturbations are washed out and unable to condense into structure. If there were only ordinary matter in the universe, there would not have been enough time for density perturbations to grow into the galaxies and clusters currently seen.

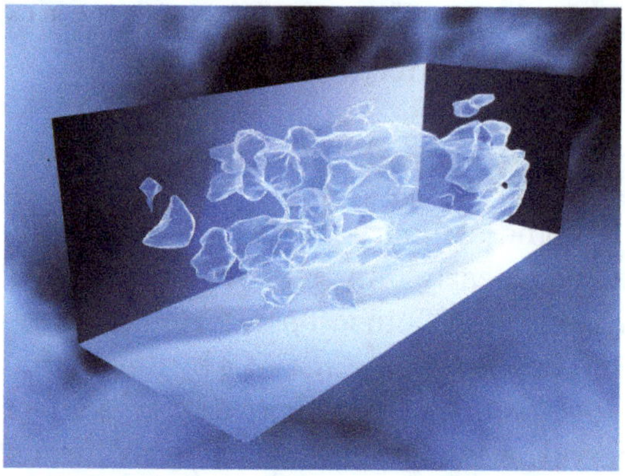

3-D map of the large-scale distribution of dark matter, reconstructed from measurements of weak gravitational lensing with the Hubble Space Telescope.

Dark matter provides a solution to this problem because it is unaffected by radiation. Therefore, its density perturbations can grow first. The resulting gravitational potential acts as an attractive potential well for ordinary matter collapsing later, speeding up the structure formation process.

Bullet Cluster

If dark matter does not exist, then the next most likely explanation must be general relativity — the prevailing theory of gravity — is incorrect and should be modified. The Bullet Cluster, the result of a recent collision of two galaxy clusters, provides a challenge for modified gravity theories because its apparent center of mass is far displaced from the baryonic center of mass. Standard dark matter models can easily explain this observation, but modified gravity has a much harder time, especially since the observational evidence is model-independent.

Type Ia Supernova Distance Measurements

Type Ia supernovae can be used as standard candles to measure extragalactic distances, which can in turn be used to measure how fast the universe has expanded in the past. Data indicates the universe is expanding at an accelerating rate, the cause of which is usually ascribed to dark energy. Since observations indicate the universe is almost flat, it is expected the total energy density of everything in the universe should sum to 1 ($\Omega_{tot} \approx 1$). The measured dark energy density is $\Omega_\Lambda \approx 0.690$; the observed ordinary (baryonic) matter energy density is $\Omega_b \approx 0.0482$ and the energy density of radiation is negligible. This leaves a missing $\Omega_{dm} \approx 0.258$ which nonetheless behaves like matter — dark matter.

Sky Surveys and Baryon Acoustic Oscillations

Baryon acoustic oscillations (BAO) are fluctuations in the density of the visible baryonic matter (normal matter) of the universe on large scales. These are predicted to arise in the Lambda-CDM model due to acoustic oscillations in the photon-baryon fluid of the early universe, and can be observed in the cosmic microwave background angular power spectrum. BAOs set up a preferred length scale for baryons. As the dark matter and baryons clumped together after recombination, the effect is much weaker in the galaxy distribution in the nearby universe, but is detectable as a subtle (≈1 percent) preference for pairs of galaxies to be separated by 147 Mpc, compared to those separated by 130–160 Mpc. This feature was predicted theoretically in the 1990s and then discovered in 2005, in two large galaxy redshift surveys, the Sloan Digital Sky Survey and the 2dF Galaxy Redshift Survey. Combining the CMB observations with BAO measurements from galaxy redshift surveys provides a precise estimate of the Hubble constant and the average matter density in the Universe. The results support the Lambda-CDM model.

Redshift-space Distortions

Large galaxy redshift surveys may be used to make a three-dimensional map of the galaxy distribution. These maps are slightly distorted because distances are estimated from observed redshifts; the redshift contains a contribution from the galaxy's so-called peculiar velocity in addition to the dominant Hubble expansion term. On average, superclusters are expanding more slowly than the cosmic mean due to their gravity, while voids are expanding faster than average. In a redshift map, galaxies in front of a supercluster have excess radial velocities towards it and have redshifts slightly higher than their distance would imply, while galaxies behind the supercluster have redshifts slightly low for their distance. This effect causes superclusters to appear squashed in the radial direction, and likewise voids are stretched. Their angular positions are unaffected. This effect is not detectable for any one structure since the true shape is not known, but can be measured by averaging over many structures. It was predicted quantitatively by Nick Kaiser in 1987, and first decisively measured in 2001 by the 2dF Galaxy Redshift Survey. Results are in agreement with the Lambda-CDM model.

Lyman-alpha Forest

In astronomical spectroscopy, the Lyman-alpha forest is the sum of the absorption lines arising from the Lyman-alpha transition of neutral hydrogen in the spectra of distant galaxies and quasars. Lyman-alpha forest observations can also constrain cosmological models. These constraints agree with those obtained from WMAP data.

Composition of Dark Matter: Baryonic vs. Nonbaryonic

There are various hypotheses about what dark matter could consist of, as set out in the table.

Some dark matter hypotheses	
Light bosons	Quantum chromodynamics axions
	Axion-like particles
	Fuzzy cold dark matter

Neutrinos	Standard Model
	Sterile neutrinos
Weak scale	Supersymmetry
	Extra dimensions
	Little Higgs
	Effective field theory
	Simplified models
Other particles	Weakly interacting massive particles
	Self-interacting dark matter
	Superfluid vacuum theory
Macroscopic	Primordial black holes
	Massive compact halo objects (MaCHOs)
	Macroscopic dark matter (Macros)
Modified gravity (MOG)	Modified Newtonian dynamics (MoND)
	Tensor–vector–scalar gravity (TeVeS)
	Entropic gravity

Dark matter can refer to any substance which interacts predominantly via gravity with visible matter (e.g., stars and planets). Hence in principle it need not be composed of a new type of fundamental particle but could, at least in part, be made up of standard baryonic matter, such as protons or neutrons. However, for the reasons outlined below, most scientists think the dark matter is dominated by a non-baryonic component, which is likely composed of a currently unknown fundamental particle (or similar exotic state).

Baryonic Matter

Baryons (protons and neutrons) make up ordinary stars and planets. However, baryonic matter also encompasses less common non-primordial black holes, neutron stars, faint old white dwarfs and brown dwarfs, collectively known as massive compact halo objects (MACHOs), which can be hard to detect.

However, multiple lines of evidence suggest the majority of dark matter is not made of baryons:

- Sufficient diffuse, baryonic gas or dust would be visible when backlit by stars.
- The theory of Big Bang nucleosynthesis predicts the observed abundance of the chemical elements. If there are more baryons, then there should also be more helium, lithium and heavier elements synthesized during the Big Bang. Agreement with observed abundances requires that baryonic matter makes up between 4–5% of the universe's critical density. In contrast, large-scale structure and other observations indicate that the total matter density is about 30% of the critical density.

- Astronomical searches for gravitational microlensing in the Milky Way found at most only a small fraction of the dark matter may be in dark, compact, conventional objects (MACHOs, etc.); the excluded range of object masses is from half the Earth's mass up to 30 solar masses, which covers nearly all the plausible candidates.
- Detailed analysis of the small irregularities (anisotropies) in the cosmic microwave background. Observations by WMAP and Planck indicate that around five sixths of the total matter is in a form that interacts significantly with ordinary matter or photons only through gravitational effects.

Non-baryonic Matter

Candidates for non-baryonic dark matter are hypothetical particles such as axions, sterile neutrinos, weakly interacting massive particles (WIMPs), gravitationally-interacting massive particles (GIMPs), supersymmetric particles, or primordial black holes. The three neutrino types already observed are indeed abundant, and dark, and matter, but because their individual masses – however uncertain they may be – are almost certainly too tiny, they can only supply a small fraction of dark matter, due to limits derived from large-scale structure and high-redshift galaxies.

Unlike baryonic matter, nonbaryonic matter did not contribute to the formation of the elements in the early universe (Big Bang nucleosynthesis) and so its presence is revealed only via its gravitational effects, or weak lensing. In addition, if the particles of which it is composed are supersymmetric, they can undergo annihilation interactions with themselves, possibly resulting in observable by-products such as gamma rays and neutrinos (indirect detection).

Dark Matter Aggregation and Dense Dark Matter Objects

If dark matter is composed of weakly-interacting particles, an obvious question is whether it can form objects equivalent to planets, stars, or black holes. Historically, the answer has been it cannot, because of two factors:

- It lacks an efficient means to lose energy: Ordinary matter forms dense objects because it has numerous ways to lose energy. Losing energy would be essential for object formation, because a particle that gains energy during compaction or falling "inward" under gravity, and cannot lose it any other way, will heat up and increase velocity and momentum. Dark matter appears to lack means to lose energy, simply because it is not capable of interacting strongly in other ways except through gravity. The virial theorem suggests that such a particle would not stay bound to the gradually forming object – as the object began to form and compact, the dark matter particles within it would speed up and tend to escape.
- It lacks a range of interactions needed to form structures: Ordinary matter

interacts in many different ways. This allows the matter to form more complex structures. For example, stars form through gravity, but the particles within them interact and can emit energy in the form of neutrinos and electromagnetic radiation through fusion when they become energetic enough. Protons and neutrons can bind via the strong interaction and then form atoms with electrons largely through electromagnetic interaction. But there is no evidence that dark matter is capable of such a wide variety of interactions, since it seems to only interact through gravity (and possibly through some means no stronger than the weak interaction, although until dark matter is better understood, this is only hopeful speculation).

In 2015–2017 the idea dense dark matter was composed of primordial black holes, made a comeback following results of gravitation wave measurements which detected the merger of intermediate mass black holes. Black holes with about 30 solar masses are not predicted to form by either stellar collapse (typically less than 15 solar masses) or by the merger of black holes in galactic centers (millions or billions of solar masses). It was proposed the intermediate mass black holes causing the detected merger formed in the hot dense early phase of the universe due to denser regions collapsing. However this was later ruled out by a survey of about a thousand supernova which detected no gravitational lensing events, although about 8 would be expected if intermediate mass primordial black holes accounted for the majority of dark matter.

The possibility atom-sized primordial black holes account for a significant fraction of dark matter was ruled out by measurements of positron and electron fluxes outside the suns heliosphere by the Voyager 1 spacecraft. Tiny black holes are theorized to emit Hawking radiation. However the detected fluxes were too low and did not have the expected energy spectrum suggesting tiny primordial black holes are not widespread enough to account for dark matter. None-the-less research and theories proposing dense dark matter accounts for dark matter continue as of 2018, including approaches to dark matter cooling, and the question remains unsettled. In 2019, the lack of microlensing effects in the observation of Andromeda suggests tiny black holes do not exist.

Classification of Dark Matter: Cold, Warm or Hot

Dark matter can be divided into cold, warm, and hot categories. These categories refer to velocity rather than an actual temperature, indicating how far corresponding objects moved due to random motions in the early universe, before they slowed due to cosmic expansion – this is an important distance called the free streaming length (FSL). Primordial density fluctuations smaller than this length get washed out as particles spread from overdense to underdense regions, while larger fluctuations are unaffected; therefore this length sets a minimum scale for later structure formation.

The categories are set with respect to the size of a protogalaxy (an object that later evolves into a dwarf galaxy): Dark matter particles are classified as cold, warm, or hot according to their FSL; much smaller (cold), similar to (warm), or much larger (hot) than a protogalaxy. Mixtures of the above are also possible: a theory of mixed dark matter was popular in the mid-1990s, but was rejected following the discovery of dark energy.

Cold dark matter leads to a bottom-up formation of structure with galaxies forming first and galaxy clusters at a latter stage, while hot dark matter would result in a top-down formation scenario with large matter aggregations forming early, later fragmenting into separate galaxies; the latter is excluded by high-redshift galaxy observations.

These categories also correspond to fluctuation spectrum effects and the interval following the Big Bang at which each type became non-relativistic. Davis et al. wrote in 1985:

> "Candidate particles can be grouped into three categories on the basis of their effect on the fluctuation spectrum. If the dark matter is composed of abundant light particles which remain relativistic until shortly before recombination, then it may be termed "hot". The best candidate for hot dark matter is a neutrino. A second possibility is for the dark matter particles to interact more weakly than neutrinos, to be less abundant, and to have a mass of order 1 keV. Such particles are termed "warm dark matter", because they have lower thermal velocities than massive neutrinos. There are at present few candidate particles which fit this description. Gravitinos and photinos have been suggested. Any particles which became nonrelativistic very early, and so were able to diffuse a negligible distance, are termed "cold" dark matter (CDM). There are many candidates for CDM including supersymmetric particles."

Another approximate dividing line is warm dark matter became non-relativistic when the universe was approximately 1 year old and 1 millionth of its present size and in the radiation-dominated era (photons and neutrinos), with a photon temperature 2.7 million Kelvins. Standard physical cosmology gives the particle horizon size as 2 c t (speed of light multiplied by time) in the radiation-dominated era, thus 2 light-years. A region of this size would expand to 2 million light-years today (absent structure formation). The actual FSL is approximately 5 times the above length, since it continues to grow slowly as particle velocities decrease inversely with the scale factor after they become non-relativistic. In this example the FSL would correspond to 10 million light-years, or 3 megaparsecs, today, around the size containing an average large galaxy.

The 2.7 million K photon temperature gives a typical photon energy of 250 electron-volts, thereby setting a typical mass scale for warm dark matter: particles much more massive than this, such as GeV–TeV mass WIMPs, would become non-relativistic much earlier than one year after the Big Bang and thus have FSLs much smaller than a protogalaxy, making them cold. Conversely, much lighter particles, such as neutrinos with masses of only a few eV, have FSLs much larger than a protogalaxy, thus qualifying them as hot.

Cold Dark Matter

Cold dark matter offers the simplest explanation for most cosmological observations. It is dark matter composed of constituents with an FSL much smaller than a protogalaxy. This is the focus for dark matter research, as hot dark matter does not seem capable of supporting galaxy or galaxy cluster formation, and most particle candidates slowed early.

The constituents of cold dark matter are unknown. Possibilities range from large objects like MACHOs (such as black holes and Preon stars) or RAMBOs (such as clusters of brown dwarfs), to new particles such as WIMPs and axions.

Studies of Big Bang nucleosynthesis and gravitational lensing convinced most cosmologists that MACHOs cannot make up more than a small fraction of dark matter.

The 1997 DAMA/NaI experiment and its successor DAMA/LIBRA in 2013, claimed to directly detect dark matter particles passing through the Earth, but many researchers remain skeptical, as negative results from similar experiments seem incompatible with the DAMA results.

Many supersymmetric models offer dark matter candidates in the form of the WIMPy Lightest Supersymmetric Particle (LSP). Separately, heavy sterile neutrinos exist in non-supersymmetric extensions to the standard model which explain the small neutrino mass through the seesaw mechanism.

Warm Dark Matter

Warm dark matter comprises particles with an FSL comparable to the size of a protogalaxy. Predictions based on warm dark matter are similar to those for cold dark matter on large scales, but with less small-scale density perturbations. This reduces the predicted abundance of dwarf galaxies and may lead to lower density of dark matter in the central parts of large galaxies. Some researchers consider this a better fit to observations. A challenge for this model is the lack of particle candidates with the required mass ≈ 300 eV to 3000 eV.

No known particles can be categorized as warm dark matter. A postulated candidate is the sterile neutrino: A heavier, slower form of neutrino that does not interact through the weak force, unlike other neutrinos. Some modified gravity theories, such as scalar–tensor–vector gravity, require "warm" dark matter to make their equations work.

Hot Dark Matter

Hot dark matter consists of particles whose FSL is much larger than the size of a protogalaxy. The neutrino qualifies as such particle. They were discovered independently, long before the hunt for dark matter: they were postulated in 1930, and detected in 1956. Neutrinos' mass is less than 10^{-6} that of an electron. Neutrinos interact with normal matter only via gravity and the weak force, making them difficult to detect (the weak force only works over a small distance, thus a neutrino triggers a weak force event only if it hits a nucleus head-on). This makes them 'weakly interacting light particles' (WILPs), as opposed to WIMPs.

The three known flavours of neutrinos are the electron, muon, and tau. Their masses are slightly different. Neutrinos oscillate among the flavours as they move. It is hard to determine an exact upper bound on the collective average mass of the three neutrinos (or for any of the three individually). For example, if the average neutrino mass were over 50 eV/c^2 (less than 10^{-5} of the mass of an electron), the universe would collapse. CMB data and other methods indicate that their average mass probably does not exceed 0.3 eV/c^2. Thus, observed neutrinos cannot explain dark matter.

Because galaxy-size density fluctuations get washed out by free-streaming, hot dark matter implies the first objects that can form are huge supercluster-size pancakes, which then fragment into galaxies. Deep-field observations show instead that galaxies formed first, followed by clusters and superclusters as galaxies clump together.

Detection of Dark Matter Particles

If dark matter is made up of sub-atomic particles, then millions, possibly billions, of such particles must pass through every square centimeter of the Earth each second. Many experiments aim to test this hypothesis. Although WIMPs are popular search candidates, the Axion Dark Matter Experiment (ADMX) searches for axions. Another candidate is heavy hidden sector particles which only interact with ordinary matter via gravity.

These experiments can be divided into two classes: direct detection experiments, which search for the scattering of dark matter particles off atomic nuclei within a detector; and indirect detection, which look for the products of dark matter particle annihilations or decays.

Direct Detection

Direct detection experiments aim to observe low-energy recoils (typically a few keVs) of nuclei induced by interactions with particles of dark matter, which (in theory) are passing through the Earth. After such a recoil the nucleus will emit energy in the form of scintillation light or phonons, as they pass through sensitive detection apparatus. To do this effectively, it is crucial to maintain a low background, and so such experiments operate deep underground to reduce the interference from cosmic rays. Examples of underground laboratories with direct detection experiments include the Stawell mine, the Soudan mine, the SNOLAB underground laboratory at Sudbury, the Gran Sasso National Laboratory, the Canfranc Underground Laboratory, the Boulby Underground Laboratory, the Deep Underground Science and Engineering Laboratory and the China Jinping Underground Laboratory.

These experiments mostly use either cryogenic or noble liquid detector technologies. Cryogenic detectors operating at temperatures below 100 mK, detect the heat produced when a particle hits an atom in a crystal absorber such as germanium. Noble liquid detectors detect scintillation produced by a particle collision in liquid xenon or argon. Cryogenic detector experiments include: CDMS, CRESST, EDELWEISS, EURECA. Noble liquid experiments include ZEPLIN, XENON, DEAP, ArDM, WARP, DarkSide, PandaX, and LUX, the Large Underground Xenon experiment. Both of these techniques focus strongly on their ability to distinguish background particles (which predominantly scatter off electrons) from dark matter particles (that scatter off nuclei). Other experiments include SIMPLE and PICASSO.

Currently there has been no well-established claim of dark matter detection from a direct detection experiment, leading instead to strong upper limits on the mass and interaction cross section with nucleons of such dark matter particles. The DAMA/NaI and more recent DAMA/LIBRA experimental collaborations have detected an annual modulation in the rate of events in their detectors, which they claim is due to dark matter. This results from the expectation that as the Earth orbits the Sun, the velocity of the detector relative to the dark matter halo will vary by a small amount. This claim is so far unconfirmed and in contradiction with negative results from other experiments such as LUX, SuperCDMS and XENON100.

A special case of direct detection experiments covers those with directional sensitivity. This is a search strategy based on the motion of the Solar System around the Galactic Center. A low-pressure time projection chamber makes it possible to access information on recoiling tracks and constrain WIMP-nucleus kinematics. WIMPs coming from the direction in which the Sun travels (approximately towards Cygnus) may then be separated from background, which should be isotropic. Directional dark matter experiments include DMTPC, DRIFT, Newage and MIMAC.

Indirect Detection

Indirect detection experiments search for the products of the self-annihilation or decay of dark matter particles in outer space. For example, in regions of high dark matter density (e.g., the centre of our galaxy) two dark matter particles could annihilate to produce gamma rays or Standard Model particle-antiparticle pairs. Alternatively if the dark matter particle is unstable, it could decay into standard model (or other) particles. These processes could be detected indirectly through an excess of gamma rays, antiprotons or positrons emanating from high density regions in our galaxy or others. A major difficulty inherent in such searches is that various astrophysical sources can mimic the signal expected from dark matter, and so multiple signals are likely required for a conclusive discovery.

A few of the dark matter particles passing through the Sun or Earth may scatter off atoms and lose energy. Thus dark matter may accumulate at the center of these bodies, increasing the chance of collision/annihilation. This could produce a distinctive signal in the form of high-energy neutrinos. Such a signal would be strong indirect proof of WIMP dark matter. High-energy neutrino telescopes such as AMANDA, IceCube and ANTARES are searching for this signal. The detection by LIGO in September 2015 of gravitational waves, opens the possibility of observing dark matter in a new way, particularly if it is in the form of primordial black holes.

Collage of six cluster collisions with dark matter maps. The clusters were observed in a study of how dark matter in clusters of galaxies behaves when the clusters collide.

Many experimental searches have been undertaken to look for such emission from dark matter annihilation or decay, examples of which follow. The Energetic Gamma Ray Experiment Telescope observed more gamma rays in 2008 than expected from the Milky Way, but scientists concluded this was most likely due to incorrect estimation of the telescope's sensitivity.

The Fermi Gamma-ray Space Telescope is searching for similar gamma rays. In April 2012, an analysis of previously available data from its Large Area Telescope instrument produced statistical evidence of a 130 GeV signal in the gamma radiation coming from the center of the Milky Way. WIMP annihilation was seen as the most probable explanation.

At higher energies, ground-based gamma-ray telescopes have set limits on the annihilation of dark matter in dwarf spheroidal galaxies and in clusters of galaxies.

The PAMELA experiment (launched in 2006) detected excess positrons. They could be from dark matter annihilation or from pulsars. No excess antiprotons were observed.

In 2013 results from the Alpha Magnetic Spectrometer on the International Space Station indicated excess high-energy cosmic rays which could be due to dark matter annihilation.

Collider Searches for Dark Matter

An alternative approach to the detection of dark matter particles in nature is to produce them in a laboratory. Experiments with the Large Hadron Collider (LHC) may be able to detect dark matter particles produced in collisions of the LHC proton beams. Because a dark matter particle should have negligible interactions with normal visible matter, it may be detected indirectly as (large amounts of) missing energy and momentum that escape the detectors, provided other (non-negligible) collision products are detected. Constraints on dark matter also exist from the LEP experiment using a similar principle, but probing the interaction of dark matter particles with electrons rather than quarks. It is important to note that any discovery from collider searches must be corroborated by discoveries in the indirect or direct detection sectors to prove that the particle discovered is, in fact, dark matter.

Alternative Hypotheses

Because dark matter remains to be conclusively identified, many other hypotheses have emerged aiming to explain the observational phenomena that dark matter was conceived to explain. The most common method is to modify general relativity. General relativity is well-tested on solar system scales, but its validity on galactic or cosmological scales has not been well proven. A suitable modification to general relativity can conceivably eliminate the need for dark matter. The best-known theories of this class are MOND and its relativistic generalization tensor-vector-scalar gravity (TeVeS), f(R) gravity, negative mass dark fluid, and entropic gravity. Alternative theories abound.

A problem with alternative hypotheses is observational evidence for dark matter comes from so many independent approaches. Explaining any individual observation is possible but explaining all of them is very difficult. Nonetheless, there have been some scattered successes for alternative hypotheses, such as a 2016 test of gravitational lensing in entropic gravity.

The prevailing opinion among most astrophysicists is while modifications to general relativity can conceivably explain part of the observational evidence, there is probably enough data to conclude there must be some form of dark matter.

DARK ENERGY

In physical cosmology, dark energy is a hypothetical form of energy that permeates all of space and tends to increase the rate of expansion of the universe. It is called "dark" because it has no electric charge and does not interact with electromagnetic radiation, such as light.

If this concept is correct, dark energy will join the other main components so far established in the universe: (a) Regular ("baryonic") matter, made of electrons and quarks;

(b) dark matter, which outweighs regular matter but is of unknown composition; and (c) relic photons (the cosmic microwave background radiation) and neutrinos from the Big Bang. (While outnumbering both regular and dark matter, these photons are thought to be so attenuated by time that their role is now minimal.)

All these components of the universe have positive mass/energy. They have an attractive gravitational interaction and are slowing down the expansion of the universe. The putative dark energy, however, has a negative mass/energy and a repelling gravitational interaction. The effect of dark energy is opposite to the positive energy components: It is speeding up the expansion of the universe.

Postulating the existence of dark energy is the most popular way to explain recent observations that the universe appears to be expanding at an accelerating rate. In the standard model of cosmology, dark energy currently accounts for almost three-quarters of the total mass-energy of the universe.

Dark energy has been used as a crucial ingredient in a recent attempt to formulate a cyclic model for the universe.

Two Proposed Forms for Dark Energy

Two proposed forms for dark energy are the cosmological constant, a constant energy density filling space homogeneously, and scalar fields such as quintessence or moduli, dynamic fields whose energy density can vary in time and space. In fact, contributions from scalar fields that are constant in space are usually also included in the cosmological constant. The cosmological constant is thought to arise from the vacuum energy. Scalar fields which do change in space are hard to distinguish from a cosmological constant, because the change may be extremely slow.

High-precision measurements of the expansion of the universe are required to understand how the speed of the expansion changes over time. The rate of expansion is parameterized by the cosmological equation of state. Measuring the equation of the state of dark energy is one of the biggest efforts in observational cosmology today.

Adding the cosmological constant to cosmology's standard Friedmann-Robertson-Walker metric (FLRW metric) leads to the Lambda-CDM model, which has been referred to as the "standard model" of cosmology because of its precise agreement with observations.

Negative Pressure

Strangely, dark energy causes expansion because it has strong negative pressure. A substance has positive pressure when it pushes outward on its surroundings. This is the usual situation for fluids. Negative pressure, or tension, exists when the substance instead pulls on its surroundings. A common example of negative pressure occurs when a solid is stretched to support a hanging weight.

According to the Friedmann-Lemaître-Robertson-Walker metric, which is an application of General Relativity to cosmology, the pressure within a substance contributes to its gravitational attraction for other things just as its mass density does. Negative pressure causes a gravitational repulsion.

The gravitational repulsive effect of dark energy's negative pressure is greater than the gravitational attraction caused by the energy itself. At the cosmological scale, it also overwhelms all other forms of gravitational attraction, resulting in the accelerating expansion of the universe.

One might wonder, how can pushing cause attraction? How can pulling cause repulsion? This sounds like a contradiction. The solution is:

- The pushing of positive pressure (and the pulling of negative pressure) are non-gravitational forces which just move substances around within space without changing space itself.
- But the gravitational attraction (or repulsion) they cause operates on space itself, decreasing (or increasing) the amount of space between things. It is this which determines the size of the universe.
- There is no necessity that these two effects should act in the same direction. In fact, they act in opposite directions.

Evidence for Dark Energy

In 1998, observations of type Ia supernovae ("one-A") by the Supernova Cosmology Project at the Lawrence Berkeley National Laboratory and the High-z Supernova Search Team suggested that the expansion of the universe is accelerating. Since then, these observations have been corroborated by several independent sources. Measurements of the cosmic microwave background, gravitational lensing, and the large scale structure of the cosmos as well as improved measurements of supernovae have been consistent with the Lambda-CDM model.

The type Ia supernovae provide the most direct evidence for dark energy. Measuring the scale factor at the time that light was emitted from an object is accomplished easily by measuring the redshift of the receding object. Finding the distance to an object is a more difficult problem, however. It is necessary to find standard candles: Objects for which the actual brightness, what astronomers call the absolute magnitude, is known, so that it is possible to relate the observed brightness, or apparent magnitude, to the distance. Without standard candles, it is impossible to measure the redshift-distance relation of Hubble's law. Type Ia supernovae are the best known standard candles for cosmological observation because they are very bright and thus visible across billions of light years. The consistency in absolute magnitude for type Ia supernovae is explained by the favored model of an old white dwarf star which gains mass from a companion star and grows until it reaches the precisely defined Chandrasekhar limit. At this mass, the white dwarf is unstable to thermonuclear runaway and explodes as a type Ia supernova with a characteristic brightness. The observed brightness of the supernovae are plotted against their redshifts, and this is used to measure the expansion history of the universe. These observations indicate that the expansion of the universe is not decelerating, which would be expected for a matter-dominated universe, but rather is mysteriously accelerating. These observations are explained by postulating a kind of energy with negative pressure: Dark energy.

The existence of dark energy, in whatever form, is needed to reconcile the measured geometry of space with the total amount of matter in the universe. Measurements of the cosmic microwave background (CMB), most recently by the WMAP satellite, indicate that the universe is very close to flat. For the shape of the universe to be flat, the mass/energy density of the Universe must be equal to a certain critical density. The total amount of matter in the Universe (including baryons and dark matter), as measured by the CMB, accounts for only about 30 percent of the critical density. This implies the existence of an additional form of energy to account for the remaining 70 percent.

The theory of large scale structure, which governs the formation of structure in the universe (stars, quasars, galaxies and galaxy clusters), also suggests that the density of matter in the universe is only 30 percent of the critical density.

The most recent WMAP observations are consistent with a Universe made up of 74 percent dark energy, 22 percent dark matter, and 4 percent ordinary matter.

Nature of Dark Energy

The exact nature of this dark energy is a matter of speculation. It is known to be very homogeneous, not very dense and is not known to interact through any of the fundamental forces other than gravity. Since it is not very dense—roughly 10^{-29} grams per cubic centimeter—it is hard to imagine experiments to detect it in the laboratory. Dark energy can only have such a profound impact on the universe, making up 70 percent of all energy, because it uniformly fills otherwise empty space. The two leading models are quintessence and the cosmological constant.

Cosmological Constant

The simplest explanation for dark energy is that it is simply the "cost of having space:" That is, a volume of space has some intrinsic, fundamental energy. This is the cosmological constant, sometimes called Lambda (hence Lambda-CDM model) after the

Greek letter Λ, the symbol used to mathematically represent this quantity. Since energy and mass are related by $E=mc^2$, Einstein's theory of general relativity predicts that it will have a gravitational effect. It is sometimes called a vacuum energy because it is the energy density of empty vacuum. In fact, most theories of particle physics predict vacuum fluctuations that would give the vacuum exactly this sort of energy. The cosmological constant is estimated by cosmologists to be on the order of 10^{-29}g/cm^3, or about 10^{-120} in reduced Planck units.

The cosmological constant has negative pressure equal to its energy density and so causes the expansion of the universe to accelerate. The reason why a cosmological constant has negative pressure can be seen from classical thermodynamics; Energy must be lost from inside a container to do work on the container. A change in volume dV requires work done equal to a change of energy −p dV, where p is the pressure. But the amount of energy in a box of vacuum energy actually increases when the volume increases (dV is positive), because the energy is equal to ρV, where ρ (rho) is the energy density of the cosmological constant. Therefore, p is negative and, in fact, p = −ρ.

A major outstanding problem is that most quantum field theories predict a huge cosmological constant from the energy of the quantum vacuum, up to 120 orders of magnitude too large. This would need to be canceled almost, but not exactly, by an equally large term of the opposite sign. Some supersymmetric theories require a cosmological constant that is exactly zero, which does not help. The present scientific consensus amounts to extrapolating the empirical evidence where it is relevant to predictions, and fine-tuning theories until a more elegant solution is found. Philosophically, the most elegant solution may be to say that if things were different, humans would not be around to observe anything—the anthropic principle. Technically, this amounts to checking theories against macroscopic observations. Unfortunately, as the known error margin in the constant predicts the fate of the universe more than its present state, many such "deeper" answers remain unknown.

Another problem arises with inclusion of the cosmic constant in the standard model, which is appearance of solutions with regions of discontinuities at low matter density. The discontinuity also affects the past sign of the vacuum energy, changing from the current negative pressure to attractive, as one looks back towards the early Universe. This finding should be considered a shortcoming of the standard model, but only when a term for vacuum energy is included.

In spite of its problems, the cosmological constant is in many respects the most economical solution to the problem of cosmic acceleration. One number successfully explains a multitude of observations. Thus, the current standard model of cosmology, the Lambda-CDM model, includes the cosmological constant as an essential feature.

Quintessence

Dark energy may become dark matter when buffeted by baryonic particles, thus leading to particle-like excitations in some type of dynamical field, referred to as quintessence. Quintessence differs from the cosmological constant in that it can vary in space and time. In order for it not to clump and form structure like matter, it must be very light so that it has a large Compton wavelength.

No evidence of quintessence is yet available, but it has not been ruled out either. It generally predicts a slightly slower acceleration of the expansion of the universe than the cosmological constant. Some scientists think that the best evidence for quintessence would come from violations of Einstein's equivalence principle and variation of the fundamental constants in space or time. Scalar fields are predicted by the standard model and string theory, but an analogous problem to the cosmological constant problem (or the problem of constructing models of cosmic inflation) occurs: Renormalization theory predicts that scalar fields should acquire large masses.

The cosmic coincidence problem asks why the cosmic acceleration began when it did. If cosmic acceleration began earlier in the universe, structures such as galaxies would never have had time to form and life, at least as it is known, would never have had a chance to exist. Proponents of the anthropic principle view this as support for their arguments. However, many models of quintessence have a so-called tracker behavior, which solves this problem. In these models, the quintessence field has a density which closely tracks (but is less than) the radiation density until matter-radiation equality, which triggers quintessence to start behaving as dark energy, eventually dominating the universe. This naturally sets the low energy scale of the dark energy.

Some special cases of quintessence are phantom energy, in which the energy density of quintessence actually increases with time, and k-essence (short for kinetic quintessence) which has a non-standard form of kinetic energy. They can have unusual properties: Phantom energy, for example, can cause a Big Rip.

Some theorists think that dark energy and cosmic acceleration are a failure of general relativity on very large scales, larger than superclusters. It is a tremendous extrapolation to think that the law of gravity, which works so well in the solar system, should work without correction on the scale of the universe. Most attempts at modifying general relativity, however, have turned out to be either equivalent to theories of quintessence, or inconsistent with observations.

Alternative ideas for dark energy have come from string theory, brane cosmology, and the holographic principle, but have not yet proved as compelling as quintessence and the cosmological constant.

Yet another, "radically conservative" class of proposals aims to explain the observational data by a more refined use of established theories rather than through the introduction of dark energy, focusing, for example, on the gravitational effects of density inhomogeneities (assumed negligible in the standard Friedmann-Robertson-Walker approximation and confirmed negligible by studies of the anisotropies of the cosmic microwave background and statistics of large-scale structure) or on consequences of electroweak symmetry breaking in the early universe.

Implications for the Fate of the Universe

Cosmologists estimate that the acceleration began roughly 9 billion years ago. Before that, it is thought that the expansion was decelerating, due to the attractive influence of dark matter and baryons. The density of dark matter in an expanding universe decreases more quickly than dark energy, and eventually the dark energy dominates. Specifically, when the volume of the universe doubles, the density of dark matter is halved but the density of dark energy is nearly unchanged (it is exactly constant in the case of a cosmological constant).

If the acceleration continues indefinitely, the ultimate result will be that galaxies outside the local supercluster will move beyond the cosmic horizon: They will no longer be visible, because their line-of-sight velocity becomes greater than the speed of light. This is not a violation of special relativity, and the effect cannot be used to send a signal between them. (Actually there is no way to even define "relative speed" in a curved spacetime. Relative speed and velocity can only be meaningfully defined in flat spacetime or in sufficiently small, infinitesimal regions of curved spacetime). Rather, it prevents any communication between them and the objects pass out of contact.

The night sky would remain the same however, to the naked eye. The Earth, the Milky Way, and the Virgo supercluster would remain virtually undisturbed while the rest of the universe recedes. In this scenario, the local supercluster would ultimately suffer heat death, just as was thought for the flat, matter-dominated universe, before measurements of cosmic acceleration.

There are some very speculative ideas about the future of the universe. One suggests that phantom energy causes divergent expansion, which would imply that the effective force of dark energy continues growing until it dominates all other forces in the universe. Under this scenario, dark energy would ultimately tear apart all gravitationally bound structures, including galaxies and solar systems, and eventually overcome the electrical and nuclear forces to tear apart atoms themselves, ending the universe in a "Big Rip." On the other hand, dark energy might dissipate with time, or even become attractive. Such uncertainties leave open the possibility that gravity might yet rule the day and lead to a universe that contracts in on itself in a "Big Crunch." Some scenarios, such as the cyclic model suggest this could be the case. While these ideas are not supported by observations, they are not ruled out. Measurements of acceleration are crucial to determining the ultimate fate of the universe in big bang theory.

References

- Vilenkin, alex (2007). Many worlds in one : the search for other universes. New york: hill and wang, a division of farrar, straus and giroux. P. 19. Isbn 978-0-8090-6722-0

- *What-is-cosmology: einsteins-theory-of-relativity-4engineers.com*, Retrieved 15 May, 2020
- *Staff (17 march 2014)*. "bicep2 2014 results release". National science foundation. Retrieved 18 march 2014
- *Epoch, science, ast: haystack.mit.edu*, Retrieved 19 May, 2020
- *Coc, a (2017). "primordial nucleosynthesis". Journal of physics: conference series. 665: 012001*. Arxiv:1609.06048. Doi:10.1088/1742-6596/665/1/012001
- *Dark-energy, entry: newworldencyclopedia.org*, Retrieved 21 March, 2020
- Overbye, dennis (20 february 2017). "cosmos controversy: the universe is expanding, but how fast?". The new york times. Retrieved 21 february 2017

Major Astronomical Tools

5 CHAPTER

Major tools that are used in astronomy are astronomical telescopes, radio telescopes, astronomical clock, visible-light detector, etc. The topics elaborated in this chapter will help in gaining a better perspective of these astronomical tools.

ASTRONOMICAL TELESCOPE

An astronomical telescope is an optical instrument which is used to see the magnified image of distant heavenly bodies like stars, planets, satellites and galaxies etc.

The final image formed by an astronomical telescope is always virtual, inverted and magnified.

Principle of Astronomical Telescope

An astronomical telescope works on the principle that when an object to be magnified is placed at a large distance from the objective lens of telescope, a virtual, inverted and magnified image of the object is formed at the least distance of distinct vision from the eye held close to the eye piece.

Construction of Astronomical Telescope

An astronomical telescope consists of two convex lenses : an objective lens O and an eye piece E. the focal length fo of the objective lens of astronomical telescope is large as compared to the focal length fe of the eye piece. And the aperture of objective lens O is large as compared to that of eye piece, so that it can receive more light from the distant object and form a bright image of the distant object. Both the objective lens and the eye piece are fitted at the free ends of two sliding tubes, at a suitable distance from each other.

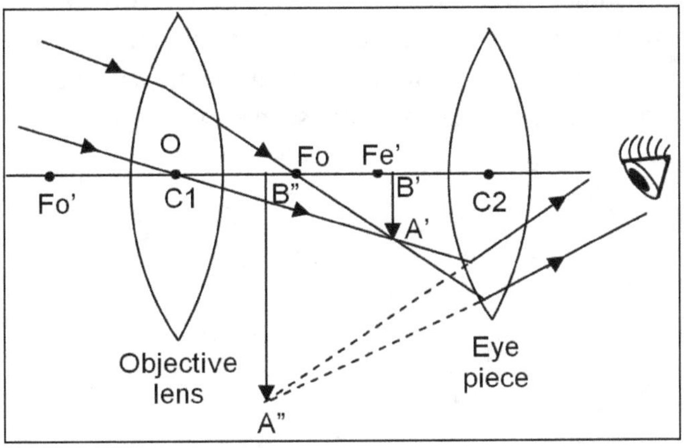

Working of Astronomical Telescope

The ray diagram to show the working of the astronomical telescope is shown in figure. A parallel beam of light from a heavenly body such as stars, planets or satellites fall on the objective lens of the telescope. The objective lens forms a real, inverted and diminished image A'B' of the heavenly body. This image (A'B') now acts as an object for the eye piece E, whose position is adjusted so that the image lies between the focus fe' and the optical centre C_2 of the eye piece. Now the eye piece forms a virtual, inverted and highly magnified image of object at infinity. When the final image of an object is formed at infinity, the telescope is said to be in 'normal adjustment'.

It should be noted that, the final image of object (such as stars, planets or satellites) formed by an astronomical telescope is always inverted with respect to the object. But it does not matter whether the image formed by an astronomical telescope is inverted or not, as all the heavenly bodies are usually spherical is shape.

Magnifying Power of an Astronomical Telescope

The Magnifying Power of a telescope is given by:

$$m = \frac{f_o}{f_e}$$

Where, f_o = Focal length of the objective lens

f_e = Focal length of the eye-piece lens

And the length (L) of the tube of telescope is equal to the sum of the focal lengths of the objective lens and the eye piece. Thus,

$L = f_o + f_e$.

RADIO TELESCOPES

In addition to visible and infrared radiation, radio waves from astronomical objects can also be detected from the surface of Earth. In the early 1930s, Karl G. Jansky, an engineer at Bell Telephone Laboratories, was experimenting with antennas for long-range radio communication when he encountered some mysterious static—radio radiation coming from an unknown source.

First Radio Telescope. This rotating radio antenna was used by Jansky in his serendipitous discovery of radio radiation from the Milky Way.

He discovered that this radiation came in strongest about four minutes earlier on each successive day and correctly concluded that since Earth's sidereal rotation period (how long it takes us to rotate relative to the stars) is four minutes shorter than a solar day, the radiation must be originating from some region fixed on the celestial sphere. Subsequent investigation showed that the source of this radiation was part of the Milky Way Galaxy; Jansky had discovered the first source of cosmic radio waves.

In 1936, Grote Reber, who was an amateur astronomer interested in radio communications, used galvanized iron and wood to build the first antenna specifically designed to receive cosmic radio waves. Over the years, Reber built several such antennas and used them to carry out pioneering surveys of the sky for celestial radio sources; he remained active in radio astronomy for more than 30 years. During the first decade, he worked practically alone because professional astronomers had not yet recognized the vast potential of radio astronomy.

Detection of Radio Energy from Space

It is important to understand that radio waves cannot be "heard": they are not the sound waves you hear coming out of the radio receiver in your home or car. Like light, radio waves are a form of electromagnetic radiation, but unlike light, we cannot detect them with our senses—we must rely on electronic equipment to pick them up. In commercial radio broadcasting, we encode sound information (music or a newscaster's voice) into radio waves. These must be decoded at the other end and then turned back into sound by speakers or headphones.

False color radio image of galaxy Cygnus A. This image shows two huge, diffuse clouds (lobes) of hot gas on either side of the galaxy. Thin jets of material are also seen, one on each side, connecting the galaxy to the lobes. The lobes are shown in three colors corresponding to the intensity of the radio energy detected. Blue is least intense and is concentrated in the regions of the lobes closest to the galaxy. Green is next and is located near the center and far edges of the lobes. Finally red is the most intense and is found at the edges of the lobes farthest from the galaxy.

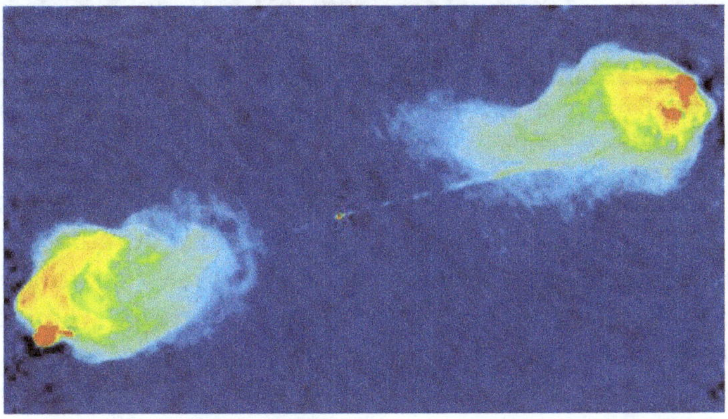

The above figure is a Radio Image. This image has been constructed of radio observations at the Very Large Array of a galaxy called Cygnus A. Colors have been added to help the eye sort out regions of different radio intensities. Red regions are the most intense, blue the least. The visible galaxy would be a small dot in the center of the image. The radio image reveals jets of expelled material (more than 160,000 light-years long) on either side of the galaxy.

The radio waves we receive from space do not, of course, have music or other program information encoded in them. If cosmic radio signals were translated into sound, they would sound like the static you hear when scanning between stations. Nevertheless, there is information in the radio waves we receive—information that can tell us about the chemistry and physical conditions of the sources of the waves.

Just as vibrating charged particles can produce electromagnetic waves, electromagnetic waves can make charged particles move back and forth. Radio waves can produce a current in conductors of electricity such as metals. An antenna is such a conductor: it intercepts radio waves, which create a feeble current in it. The current is then amplified in a radio receiver until it is strong enough to measure or record. Like your television or radio, receivers can be tuned to select a single frequency (channel). In astronomy, however, it is more common to use sophisticated data-processing techniques that allow thousands of separate frequency bands to be detected simultaneously. Thus, the astronomical radio receiver operates much like a spectrometer on a visible-light or infrared telescope, providing information about how much radiation we receive at each wavelength or frequency. After computer processing, the radio signals are recorded on magnetic disks for further analysis.

Robert C. Byrd Green Bank Telescope: This fully steerable radio telescope in West Virginia went into operation in August 2000. Its dish is about 100 meters across.

Radio waves are reflected by conducting surfaces, just as light is reflected from a shiny metallic surface, and according to the same laws of optics. A radio-reflecting telescope consists of a concave metal reflector (called a dish), analogous to a telescope mirror. The radio waves collected by the dish are reflected to a focus, where they can then be directed to a receiver and analyzed. Because humans are such visual creatures, radio astronomers often construct a pictorial representation of the radio sources they observe. Figure shows such a radio image of a distant galaxy, where radio telescopes reveal vast jets and complicated regions of radio emissions that are completely invisible in photographs taken with light.

Radio astronomy is a young field compared with visible-light astronomy, but it has experienced tremendous growth in recent decades. The world's largest radio reflectors that can be pointed to any direction in the sky have apertures of 100 meters. One of these has been built at the US National Radio Astronomy Observatory in West Virginia.

Table: Lists some of the major radio telescopes of the world.

Major Radio Observatories of the World		
Observatory	Location	Description
Individual Radio Dishes		
Arecibo Observatory	Arecibo, Puerto Rico	305-m fixed dish
Green Bank Telescope (GBT)	Green Bank, WV	110 × 100-m steerable dish
Effelsberg 100-m Telescope	Bonn, Germany	100-m steerable dish
Lovell Telescope	Manchester, England	76-m steerable dish
Canberra Deep Space Communication Complex (CDSCC)	Tidbinbilla, Australia	70-m steerable dish

Goldstone Deep Space Communications Complex (GDSCC)	Barstow, CA	70-m steerable dish
Parkes Observatory	Parkes, Australia	64-m steerable dish
Arrays of Radio Dishes		
Square Kilometre Array (SKA)	South Africa and Western Australia	Thousands of dishes, km^2 collecting area, partial array in 2020
Atacama Large Millimeter/submillimeter Array (ALMA)	Atacama desert, Northern Chile	66 7-m and 12-m dishes
Very Large Array (VLA)	Socorro, New Mexico	27-element array of 25-m dishes (36-km baseline)
Westerbork Synthesis Radio Telescope (WSRT)	Westerbork, the Netherlands	12-element array of 25-m dishes (1.6-km baseline)
Very Long Baseline Array (VLBA)	Ten US sites, HI to the Virgin Islands	10-element array of 25-m dishes (9000 km baseline)
Australia Telescope Compact Array (ATCA)	Several sites in Australia	8-element array (seven 22-m dishes plus Parkes 64 m)
Multi-Element Radio Linked Interferometer Network (MERLIN)	Cambridge, England, and other British sites	Network of seven dishes (the largest is 32 m)
Millimeter-wave Telescopes		
IRAM	Granada, Spain	30-m steerable mm-wave dish
James Clerk Maxwell Telescope (JCMT)	Mauna Kea, HI	15-m steerable mm-wave dish
Nobeyama Radio Observatory (NRO)	Minamimaki, Japan	6-element array of 10-m wave dishes
Hat Creek Radio Observatory (HCRO)	Cassel, CA	6-element array of 5-m wave dishes

Radio Interferometry

As we discussed earlier, a telescope's ability to show us fine detail (its resolution) depends upon its aperture, but it also depends upon the wavelength of the radiation that the telescope is gathering. The longer the waves, the harder it is to resolve fine detail in the images or maps we make. Because radio waves have such long wavelengths, they present tremendous challenges for astronomers who need good resolution. In fact, even the largest radio dishes on Earth, operating alone, cannot make out as much detail as the typical small visible-light telescope used in a college astronomy lab. To overcome this difficulty, radio astronomers have learned to sharpen their images by linking two or more radio telescopes together electronically. Two or more telescopes linked together in this way are called an interferometer.

"Interferometer" may seem like a strange term because the telescopes in an interferometer work cooperatively; they don't "interfere" with each other. Interference, however, is a technical term for the way that multiple waves interact with each other when they arrive in our instruments, and this interaction allows us to coax more detail out of our observations. The resolution of an interferometer depends upon the separation of the telescopes, not upon their individual apertures. Two telescopes separated by 1 kilometer provide the same resolution as would a single dish 1 kilometer across (although they are not, of course, able to collect as much radiation as a radio-wave bucket that is 1 kilometer across).

Atacama Large Millimeter/Submillimeter Array (ALMA): Located in the Atacama Desert of Northern Chile, ALMA currently provides the highest resolution for radio observations.

To get even better resolution, astronomers combine a large number of radio dishes into an interferometer array. In effect, such an array works like a large number of two-dish interferometers, all observing the same part of the sky together. Computer processing of the results permits the reconstruction of a high-resolution radio image. The most extensive such instrument in the United States is the National Radio Astronomy Observatory's Very Large Array (VLA) near Socorro, New Mexico. It consists of 27 movable radio telescopes (on railroad tracks), each having an aperture of 25 meters, spread over a total span of about 36 kilometers. By electronically combining the signals from all of its individual telescopes, this array permits the radio astronomer to make pictures of the sky at radio wavelengths comparable to those obtained with a visible-light telescope, with a resolution of about 1 arcsecond.

The Atacama Large Millimeter/submillimeter array (ALMA) in the Atacama Desert of Northern Chile, at an altitude of 16,400 feet, consists of 12 7-meter and 54 12-meter telescopes, and can achieve baselines up to 16 kilometers. Since it became operational in 2013, it has made observations at resolutions down to 6 milliarcseconds (0.006 arcseconds), a remarkable achievement for radio astronomy.

Initially, the size of interferometer arrays was limited by the requirement that all of the dishes be physically wired together. The maximum dimensions of the array were thus only a few tens of kilometers. However, larger interferometer separations can be achieved if the telescopes do not require a physical connection. Astronomers, with the use of current technology and computing power, have learned to time the arrival of electromagnetic waves coming from space very precisely at each telescope and combine the data later. If the telescopes are as far apart as California and Australia, or as West Virginia and Crimea in Ukraine, the resulting resolution far surpasses that of visible-light telescopes.

The United States operates the Very Long Baseline Array (VLBA), made up of 10 individual telescopes stretching from the Virgin Islands to Hawaii. The VLBA, completed in 1993, can form astronomical images with a resolution of 0.0001 arcseconds, permitting features as small as 10 astronomical units (AU) to be distinguished at the center of our Galaxy.

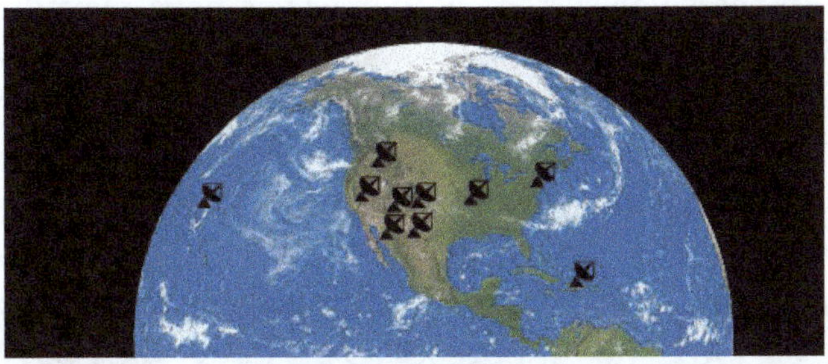

Very Long Baseline Array: This map shows the distribution of 10 antennas that constitute an array of radio telescopes stretching across the United States and its territories.

Recent advances in technology have also made it possible to do interferometry at visible-light and infrared wavelengths. At the beginning of the twenty-first century, three observatories with multiple telescopes each began using their dishes as interferometers, combining their light to obtain a much greater resolution. In addition, a dedicated interferometric array was built on Mt. Wilson in California. Just as in radio arrays, these observations allow astronomers to make out more detail than a single telescope could provide.

Visible-Light Interferometers				
Longest Baseline (m)	Telescope Name	Location	Mirrors	Status
400	CHARA Array (Center for High Angular Resolution Astronomy)	Mount Wilson, CA	Six 1-m telescopes	Operational since 2004
200	Very Large Telescope	Cerro Paranal, Chile	Four 8.2-m telescopes	Completed 2000
85	Keck I and II telescopes	Mauna Kea, HI	Two 10-m telescopes	Operated from 2001 to 2012

22.8	Large Binocular Telescope	Mount Graham, AZ	Two 8.4-m telescopes	First light 2004

Radar Astronomy

Radar is the technique of transmitting radio waves to an object in our solar system and then detecting the radio radiation that the object reflects back. The time required for the round trip can be measured electronically with great precision. Because we know the speed at which radio waves travel (the speed of light), we can determine the distance to the object or a particular feature on its surface (such as a mountain).

Radar observations have been used to determine the distances to planets and how fast things are moving in the solar system. Radar waves have played important roles in navigating spacecraft throughout the solar system. In addition, as will be discussed in later chapters, radar observations have determined the rotation periods of Venus and Mercury, probed tiny Earth-approaching asteroids, and allowed us to investigate the mountains and valleys on the surfaces of Mercury, Venus, Mars, and the large moons of Jupiter.

Largest Radio and Radar Dish: The Arecibo Observatory, with its 1000-foot radio dish-filling valley in Puerto Rico, is part of the National Astronomy and Ionosphere Center, operated by SRI International, USRA, and UMET under a cooperative agreement with the National Science Foundation.

Any radio dish can be used as a radar telescope if it is equipped with a powerful transmitter as well as a receiver. The most spectacular facility in the world for radar astronomy is the 1000-foot (305-meter) telescope at Arecibo in Puerto Rico. The Arecibo telescope is too large to be pointed directly at different parts of the sky. Instead, it is constructed in a huge natural "bowl" (more than a mere dish) formed by several hills, and it is lined with reflecting metal panels. A limited ability to track astronomical sources is achieved by moving the receiver system, which is suspended on cables 100 meters above the surface of the bowl. An even larger (500-meter) radar telescope is currently under construction. It is the Five-hundred-meter Aperture Spherical Telescope (FAST) in China and is expected to be completed in 2016.

In the 1930s, radio astronomy was pioneered by Karl G. Jansky and Grote Reber. A radio telescope is basically a radio antenna (often a large, curved dish) connected to a receiver. Significantly enhanced resolution can be obtained with interferometers, including interferometer arrays like the 27-element VLA and the 66-element ALMA. Expanding to very long baseline interferometers, radio astronomers can achieve resolutions as precise as 0.0001 arcsecond. Radar astronomy involves transmitting as well as receiving. The largest radar telescope currently in operation is a 305-meter bowl at Arecibo.

VISIBLE-LIGHT DETECTORS AND INSTRUMENTS

The first detector used for astronomical observations was the human eye, but it suffers from being connected to an imperfect recording and retrieving device—the human brain. Photography and modern electronic detectors have eliminated the quirks of human memory by making a permanent record of the information from the cosmos.

The eye also suffers from having a very short integration time; it takes only a fraction of a second to add light energy together before sending the image to the brain. One important advantage of modern detectors is that the light from astronomical objects can be collected by the detector over longer periods of time; this technique is called "taking a long exposure." Exposures of several hours are required to detect very faint objects in the cosmos.

Before the light reaches the detector, astronomers today normally use some type of instrument to sort the light according to wavelength. The instrument may be as simple as colored filters, which transmit light within a specified range of wavelengths. A red transparent plastic is an everyday example of a filter that transmits only the red light and blocks the other colors. After the light passes through a filter, it forms an image that astronomers can then use to measure the apparent brightness and color of objects. We will show you many examples of such images in the later chapters of this book, and we will describe what we can learn from them.

Alternatively, the instrument between telescope and detector may be one of several devices that spread the light out into its full rainbow of colors so that astronomers can measure individual lines in the spectrum. Such an instrument (which you learned about in the chapter on Radiation and Spectra) is called a spectrometer because it allows astronomers to measure (to meter) the spectrum of a source of radiation. Whether a filter or a spectrometer, both types of wavelength-sorting instruments still have to use detectors to record and measure the properties of light.

Photographic and Electronic Detectors

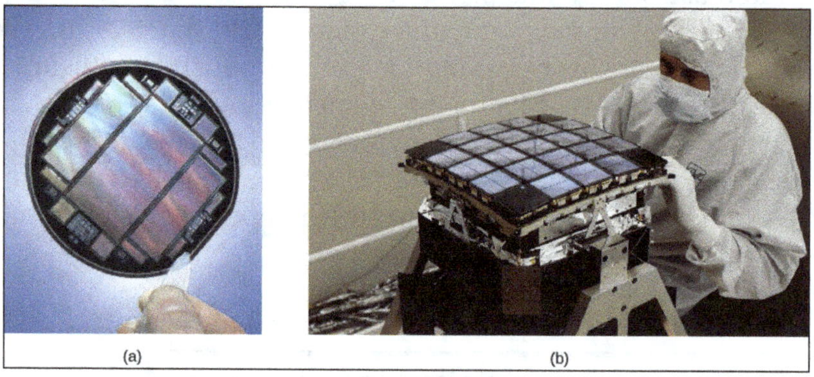

Charge-Coupled Devices (CCDs): (a) This CCD is a mere 300-micrometers thick (thinner than a human hair) yet holds more than 21 million pixels. (b) This matrix of 42 CCDs serves the Kepler telescope.

Throughout most of the twentieth century, photographic film or glass plates served as the prime astronomical detectors, whether for photographing spectra or direct images of celestial objects. In a photographic plate, a light-sensitive chemical coating is applied to a piece of glass that, when developed, provides a lasting record of the image. At observatories around the world, vast collections of photographs preserve what the sky has looked like during the past 100 years. Photography represents a huge improvement over the human eye, but it still has limitations. Photographic films are inefficient: only about 1% of the light that actually falls on the film contributes to the chemical change that makes the image; the rest is wasted.

Astronomers today have much more efficient electronic detectors to record astronomical images. Most often, these are charge-coupled devices (CCDs), which are similar to the detectors used in video camcorders or in digital cameras. In a CCD, photons of radiation hitting any part of the detector generate a stream of charged particles (electrons) that are stored and counted at the end of the exposure. Each place where the radiation is counted is called a pixel (picture element), and modern detectors can count the photons in millions of pixels (megapixels, or MPs).

Because CCds typically record as much as 60–70% of all the photons that strike them, and the best silicon and infrared CCDs exceed 90% sensitivity, we can detect much fainter objects. Among these are many small moons around the outer planets, icy dwarf planets beyond Pluto, and dwarf galaxies of stars. CCDs also provide more accurate measurements of the brightness of astronomical objects than photography, and their output is digital—in the form of numbers that can go directly into a computer for analysis.

Infrared Observations

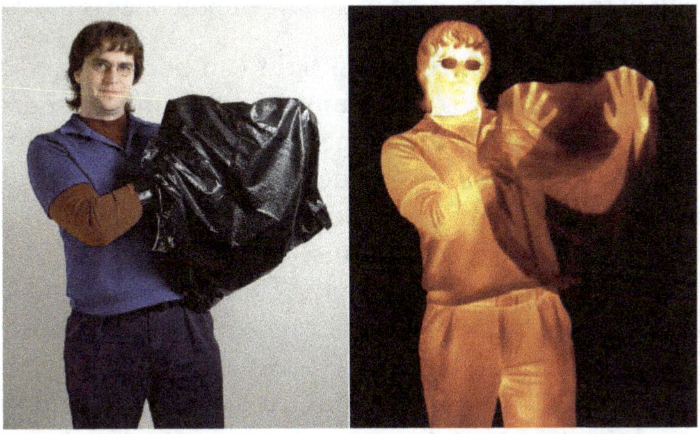

Infrared Eyes: Infrared waves can penetrate places in the universe from which light is blocked, as shown in this infrared image where the plastic bag blocks visible light but not infrared. To solve this problem, astronomers must protect the infrared detector from nearby radiation, just as you would shield photographic film from bright daylight. Since anything warm radiates infrared energy, the detector must be isolated in very cold surroundings; often, it is held near absolute zero (1 to 3 K) by immersing it in liquid helium. The second step is to reduce the radiation emitted by the telescope structure and optics and to block this heat from reaching the infrared detector.

Observing the universe in the infrared band of the spectrum presents some additional challenges. The infrared region extends from wavelengths near 1 micrometer (μm), which is about the long wavelength sensitivity limit of both CCDs and photography, to 100 micrometers or longer. Recall from the discussion on radiation and spectra that infrared is "heat radiation". The main challenge to astronomers using infrared is to distinguish between the tiny amount of heat radiation that reaches Earth from stars and galaxies, and the much greater heat radiated by the telescope itself and our planet's atmosphere.

Typical temperatures on Earth's surface are near 300 K, and the atmosphere through which observations are made is only a little cooler. The telescope, the observatory, and even the sky are radiating infrared energy with a peak wavelength of about 10 micrometers. To infrared eyes, everything on Earth is brightly aglow—including the telescope and camera. The challenge is to detect faint cosmic sources against this sea of infrared light. Another way to look at this is that an astronomer using infrared must always contend with the situation that a visible-light observer would face if working in broad daylight with a telescope and optics lined with bright fluorescent lights.

Spectroscopy

Spectroscopy is one of the astronomer's most powerful tools, providing information about the composition, temperature, motion, and other characteristics of celestial objects. More than half of the time spent on most large telescopes is used for spectroscopy.

The many different wavelengths present in light can be separated by passing them through a spectrometer to form a spectrum. Light from the source (actually, the image of a source produced by the telescope) enters the instrument through a small hole or narrow slit, and is collimated (made into a beam of parallel rays) by a lens. The light then passes through a prism, producing a spectrum: different wavelengths leave the prism in different directions because each wavelength is bent by a different amount when it enters and leaves the prism. A second lens placed behind the prism focuses the many different images of the slit or entrance hole onto a CCD or other detecting device. This collection of images (spread out by color) is the spectrum that astronomers can then analyze at a later point. As spectroscopy spreads the light out into more and more collecting bins, fewer photons go into each bin, so either a larger telescope is needed or the integration time must be greatly increased—usually both.

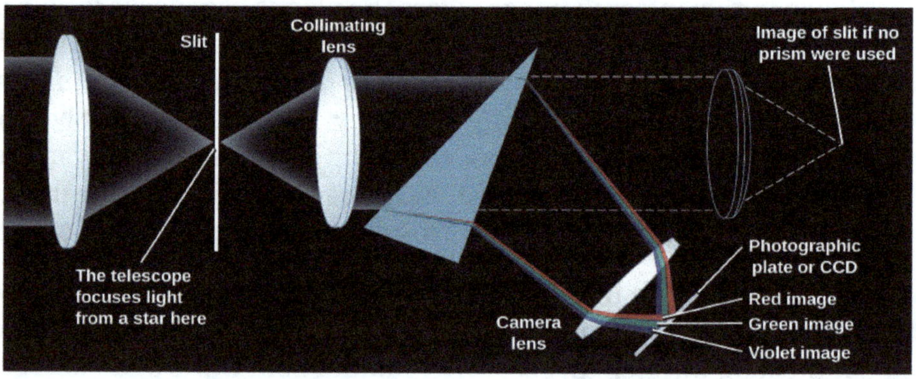

Prism Spectrometer: The light from the telescope is focused on a slit. A prism (or grating) disperses the light into a spectrum, which is then photographed or recorded electronically.

In practice, astronomers today are more likely to use a different device, called a grating, to disperse the spectrum. A grating is a piece of material with thousands of grooves on its surface. While it functions completely differently, a grating, like a prism, also spreads light out into a spectrum.

Visible-light detectors include the human eye, photographic film, and charge-coupled devices (CCDs). Detectors that are sensitive to infrared radiation must be cooled to very low temperatures since everything in and near the telescope gives off infrared waves. A spectrometer disperses the light into a spectrum to be recorded for detailed analysis.

ASTRONOMICAL SPECTROSCOPY

Astronomical spectroscopy is the study of astronomy using the techniques of spectroscopy to measure the spectrum of electromagnetic radiation, including visible light and radio, which radiates from stars and other celestial objects. A stellar spectrum can reveal many properties of stars, such as their chemical composition, temperature, density, mass, distance, luminosity, and relative motion using Doppler shift measurements. Spectroscopy is also used to study the physical properties of many other types of celestial objects such as planets, nebulae, galaxies, and active galactic nuclei.

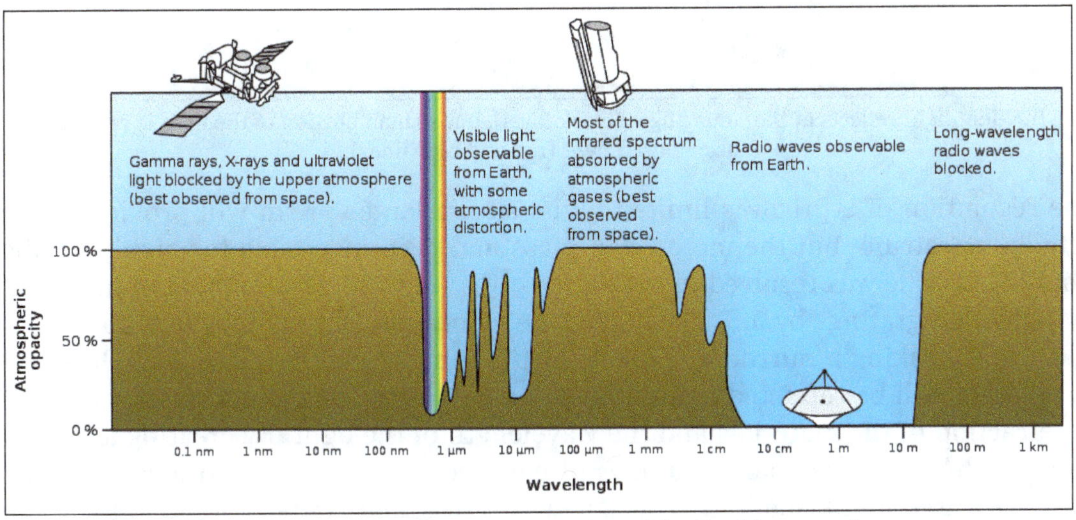

Electromagnetic transmittance, or opacity, of the Earth's atmosphere.

Astronomical spectroscopy is used to measure three major bands of radiation: visible spectrum, radio, and X-ray. While all spectroscopy looks at specific areas of the spectrum, different methods are required to acquire the signal depending on the frequency. Ozone (O_3) and molecular oxygen (O_2) absorb light with wavelengths under 300 nm, meaning that X-ray and ultraviolet spectroscopy require the use of a satellite telescope or rocket mounted detectors. Radio signals have much longer wavelengths than optical signals, and require the use of antennas or radio dishes. Infrared light is absorbed by atmospheric water and carbon dioxide, so while the equipment is similar to that used in optical spectroscopy, satellites are required to record much of the infrared spectrum.

Optical Spectroscopy

Physicists have been looking at the solar spectrum since Isaac Newton first used a simple prism to observe the refractive properties of light. In the early 1800s Joseph von Fraunhofer used his skills as a glass maker to create very pure prisms, which allowed him to observe 574 dark lines in a seemingly continuous spectrum. Soon after this, he combined telescope and prism to observe the spectrum of Venus, the Moon, Mars, and various stars such as Betelgeuse; his company continued to manufacture and sell high-quality refracting telescopes based on his original designs until its closure in 1884.

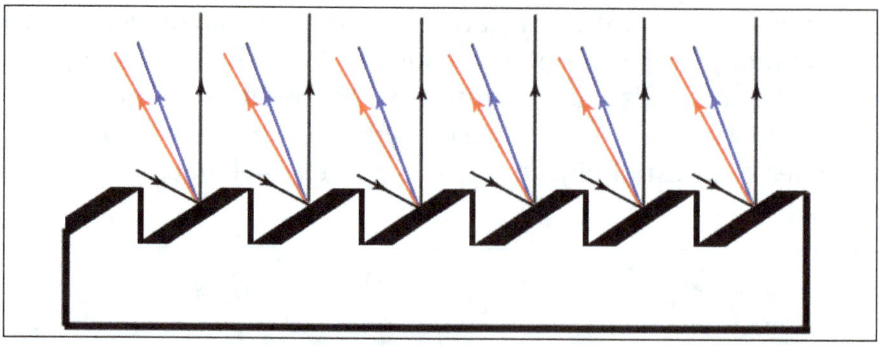

Incident light reflects at the same angle (black lines), but a small portion of the light is refracted as coloured light (red and blue lines).

The resolution of a prism is limited by its size; a larger prism will provide a more detailed spectrum, but the increase in mass makes it unsuitable for highly detailed work. This issue was resolved in the early 1900s with the development of high-quality reflection gratings by J.S. Plaskett at the Dominion Observatory in Ottawa, Canada. Light striking a mirror will reflect at the same angle, however a small portion of the light will be refracted at a different angle; this is dependent upon the indices of refraction of the materials and the wavelength of the light. By creating a "blazed" grating which utilizes a large number of parallel mirrors, the small portion of light can be focused and visualized. These new spectroscopes were more detailed than a prism, required less light, and could be focused on a specific region of the spectrum by tilting the grating.

The limitation to a blazed grating is the width of the mirrors, which can only be ground a finite amount before focus is lost; the maximum is around 1000 lines/mm. In order to overcome this limitation holographic gratings were developed. Volume phase holographic gratings use a thin film of dichromated gelatin on a glass surface, which is subsequently exposed to a wave pattern created by an interferometer. This wave pattern sets up a reflection pattern similar to the blazed gratings but utilizing Bragg diffraction, a process where the angle of reflection is dependent on the arrangement of the atoms in the gelatin. The holographic gratings can have up to 6000 lines/mm and can be up to twice as efficient in collecting light as blazed gratings. Because they are sealed between two sheets of glass, the holographic gratings are very versatile, potentially lasting decades before needing replacement.

Light dispersed by the grating or prism in a spectrograph can be recorded by a detector. Historically, photographic plates were widely used to record spectra until electronic detectors were developed, and today optical spectrographs most often employ charge-coupled devices (CCDs). The wavelength scale of a spectrum can be calibrated by observing the spectrum of emission lines of known wavelength from a gas-discharge lamp. The flux scale of a spectrum can be calibrated as a function of wavelength by comparison with an observation of a standard star with corrections for atmospheric absorption of light; this is known as spectrophotometry.

Radio Spectroscopy

Radio astronomy was founded with the work of Karl Jansky in the early 1930s, while working for Bell Labs. He built a radio antenna to look at potential sources of interference for transatlantic radio transmissions. One of the sources of noise discovered came not from Earth, but from the center of the Milky Way, in the constellation Sagittarius. In 1942, JS Hey captured the sun's radio frequency using military radar receivers. Radio spectroscopy started with the discovery of the 21-centimeter H I line in 1951.

Radio interferometry was pioneered in 1946, when Joseph Lade Pawsey, Ruby Payne-Scott and Lindsay McCready used a single antenna atop a sea cliff to observe 200 MHz solar radiation. Two incident beams, one directly from the sun and the other reflected from the sea surface, generated the necessary interference. The first multi-receiver interferometer was built in the same year by Martin Ryle and Vonberg. In 1960, Ryle and Antony Hewish published the technique of aperture synthesis to analyze interferometer data. The aperture synthesis process, which involves autocorrelating and discrete Fourier transforming the incoming signal, recovers both the spatial and frequency variation in flux. The result is a 3D image whose third axis is frequency. For this work, Ryle and Hewish were jointly awarded the 1974 Nobel Prize in Physics.

Stars and their Properties

Chemical Properties

Newton used a prism to split white light into a spectrum of color, and Fraunhofer's high-quality prisms allowed scientists to see dark lines of an unknown origin. In the 1850s, Gustav Kirchhoff and Robert Bunsen described the phenomena behind these dark lines. Hot solid objects produce light with a continuous spectrum, hot gases emit light at specific wavelengths, and hot solid objects surrounded by cooler gases show a near-continuous spectrum with dark lines corresponding to the emission lines of the gases. By comparing the absorption lines of the Sun with emission spectra of known gases, the chemical composition of stars can be determined.

The major Fraunhofer lines, and the elements with which they are associated, appear in the following table. Designations from the early Balmer Series are shown in parentheses.

Designation	Element	Wavelength (nm)	Designation	Element	Wavelength (nm)
y	O_2	898.765	c	Fe	495.761
Z	O_2	822.696	F (Hβ)	H	486.134
A	O_2	759.370	d	Fe	466.814
B	O_2	686.719	e	Fe	438.355
C (Hα)	H	656.281	G' (Hγ)	H	434.047
a	O_2	627.661	G	Fe	430.790
D_1	Na	589.592	G	Ca	430.774

D_2	Na	588.995	h (Hδ)	H	410.175
D_3 or d	He	587.5618	H	Ca^+	396.847
e	Hg	546.073	K	Ca^+	393.368
E_2	Fe	527.039	L	Fe	382.044
b_1	Mg	518.362	N	Fe	358.121
b_2	Mg	517.270	P	Ti^+	336.112
b_3	Fe	516.891	T	Fe	302.108
b_4	Mg	516.733	t	Ni	299.444

Not all of the elements in the Sun were immediately identified. Two examples are listed below.

- In 1868 Norman Lockyer and Pierre Janssen independently observed a line next to the sodium doublet (D_1 and D_2) which Lockyer determined to be a new element. He named it Helium, but it wasn't until 1895 the element was found on Earth.
- In 1869 the astronomers Charles Augustus Young and William Harkness independently observed a novel green emission line in the Sun's corona during an eclipse. This "new" element was incorrectly named coronium, as it was only found in the corona. It was not until the 1930s that Walter Grotrian and Bengt Edlén discovered that the spectral line at 530.3 nm was due to highly ionized iron (Fe^{13+}). Other unusual lines in the coronal spectrum are also caused by highly charged ions, such as nickel and calcium, the high ionization being due to the extreme temperature of the solar corona.

To date more than 20,000 absorption lines have been listed for the Sun between 293.5 and 877.0 nm, yet only approximately 75% of these lines have been linked to elemental absorption.

By analyzing the width of each spectral line in an emission spectrum, both the elements present in a star and their relative abundances can be determined. Using this information stars can be categorized into stellar populations; Population I stars are the youngest stars and have the highest metal content (our Sun is a Pop I star), while Population III stars are the oldest stars with a very low metal content.

Temperature and Size

In 1860 Gustav Kirchhoff proposed the idea of a black body, a material that emits electromagnetic radiation at all wavelengths. In 1894 Wilhelm Wien derived an expression relating the temperature (T) of a black body to its peak emission wavelength (λ_{max}).

$$\lambda_{\max} T = b$$

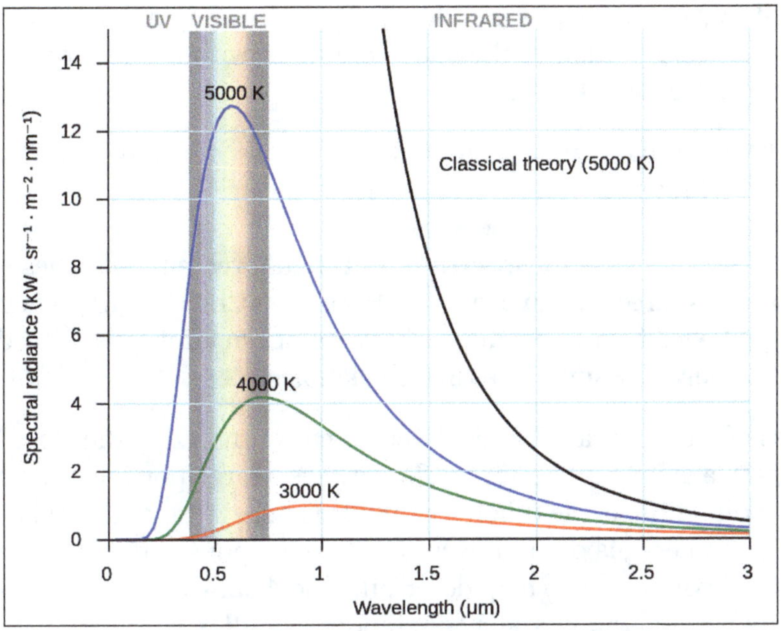

Black body curves for various temperatures.

b is a constant of proportionality called Wien's displacement constant, equal to 2.897771955...$\times 10^{-3}$ m·K. This equation is called Wien's Law. By measuring the peak wavelength of a star, the surface temperature can be determined. For example, if the peak wavelength of a star is 502 nm the corresponding temperature will be 5778 kelvins.

The luminosity of a star is a measure of the electromagnetic energy output in a given amount of time. Luminosity (L) can be related to the temperature (T) of a star by:

$$L = 4\pi R^2 \sigma T^4,$$

where R is the radius of the star and σ is the Stefan–Boltzmann constant, with a value of 5.670374419...$\times 10^{-8}$ W·m^{-2}·K^{-4}. Thus, when both luminosity and temperature are known (via direct measurement and calculation) the radius of a star can be determined.

Galaxies

The spectra of galaxies look similar to stellar spectra, as they consist of the combined light of billions of stars.

Doppler shift studies of galaxy clusters by Fritz Zwicky in 1937 found that the galaxies in a cluster were moving much faster than seemed to be possible from the mass of the cluster inferred from the visible light. Zwicky hypothesized that there must be a great deal of non-luminous matter in the galaxy clusters, which became known as dark matter. Since his discovery, astronomers have determined that a large portion of galaxies (and most of the universe) is made up of dark matter. In 2003, however, four galaxies

(NGC 821, NGC 3379, NGC 4494, and NGC 4697) were found to have little to no dark matter influencing the motion of the stars contained within them; the reason behind the lack of dark matter is unknown.

In the 1950s, strong radio sources were found to be associated with very dim, very red objects. When the first spectrum of one of these objects was taken there were absorption lines at wavelengths where none were expected. It was soon realised that what was observed was a normal galactic spectrum, but highly red shifted. These were named quasi-stellar radio sources, or quasars, by Hong-Yee Chiu in 1964. Quasars are now thought to be galaxies formed in the early years of our universe, with their extreme energy output powered by super-massive black holes.

The properties of a galaxy can also be determined by analyzing the stars found within them. NGC 4550, a galaxy in the Virgo Cluster, has a large portion of its stars rotating in the opposite direction as the other portion. It is believed that the galaxy is the combination of two smaller galaxies that were rotating in opposite directions to each other. Bright stars in galaxies can also help determine the distance to a galaxy, which may be a more accurate method than parallax or standard candles.

Interstellar Medium

The interstellar medium is matter that occupies the space between star systems in a galaxy. 99% of this matter is gaseous - hydrogen, helium, and smaller quantities of other ionized elements such as oxygen. The other 1% is dust particles, thought to be mainly graphite, silicates, and ices. Clouds of the dust and gas are referred to as nebulae.

There are three main types of nebula: absorption, reflection, and emission nebulae. Absorption (or dark) nebulae are made of dust and gas in such quantities that they obscure the starlight behind them, making photometry difficult. Reflection nebulae, as their name suggest, reflect the light of nearby stars. Their spectra are the same as the stars surrounding them, though the light is bluer; shorter wavelengths scatter better than longer wavelengths. Emission nebulae emit light at specific wavelengths depending on their chemical composition.

Gaseous Emission Nebulae

In the early years of astronomical spectroscopy, scientists were puzzled by the spectrum of gaseous nebulae. In 1864 William Huggins noticed that many nebulae showed only emission lines rather than a full spectrum like stars. From the work of Kirchhoff, he concluded that nebulae must contain "enormous masses of luminous gas or vapour." However, there were several emission lines that could not be linked to any terrestrial element, brightest among them lines at 495.9 nm and 500.7 nm. These lines were attributed to a new element, nebulium, until Ira Bowen determined in 1927 that the emission lines were from highly ionised oxygen (O^{+2}). These emission lines could not be replicated in a laboratory because they are forbidden lines; the low density of a nebula (one atom per cubic centimetre) allows for metastable ions to decay via forbidden line emission rather than collisions with other atoms.

Not all emission nebulae are found around or near stars where solar heating causes ionisation. The majority of gaseous emission nebulae are formed of neutral hydrogen. In the ground state neutral hydrogen has two possible spin states: the electron has either the same spin or the opposite spin of the proton. When the atom transitions between these two states, it releases an emission or absorption line of 21 cm. This line is within the radio range and allows for very precise measurements:

- Velocity of the cloud can be measured via Doppler shift
- The intensity of the 21 cm line gives the density and number of atoms in the cloud
- The temperature of the cloud can be calculated

Using this information the shape of the Milky Way has been determined to be a spiral galaxy, though the exact number and position of the spiral arms is the subject of ongoing research.

Complex Molecules

Dust and molecules in the interstellar medium not only obscures photometry, but also causes absorption lines in spectroscopy. Their spectral features are generated by transitions of component electrons between different energy levels, or by rotational or vibrational spectra. Detection usually occurs in radio, microwave, or infrared portions of the spectrum. The chemical reactions that form these molecules can happen in cold, diffuse clouds or in the hot ejecta around a white dwarf star from a nova or supernova. Polycyclic aromatic hydrocarbons such as acetylene (C_2H_2) generally group together to form graphites or other sooty material, but other organic molecules such as acetone ($(CH_3)_2CO$) and buckminsterfullerenes (C_{60} and C_{70}) have been discovered.

Motion in the Universe

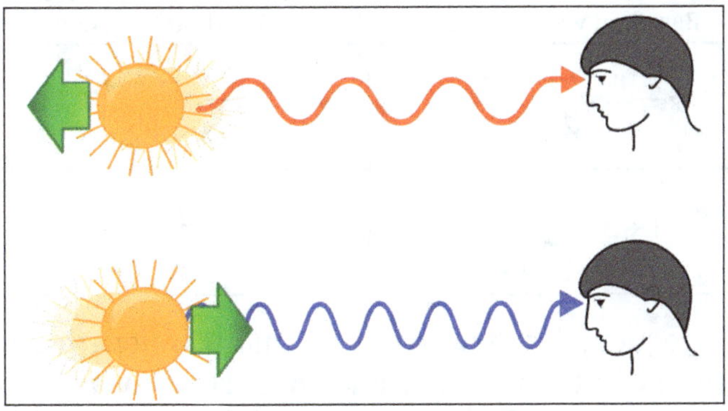

Redshift and blueshift.

Stars and interstellar gas are bound by gravity to form galaxies, and groups of galaxies can be bound by gravity in galaxy clusters. With the exception of stars in the Milky Way and the galaxies in the Local Group, almost all galaxies are moving away from us due to the expansion of the universe.

Doppler Effect and Redshift

The motion of stellar objects can be determined by looking at their spectrum. Because of the Doppler effect, objects moving towards us are blueshifted, and objects moving away are redshifted. The wavelength of redshifted light is longer, appearing redder than the source. Conversely, the wavelength of blueshifted light is shorter, appearing bluer than the source light:

$$\frac{\lambda - \lambda_0}{\lambda_0} = \frac{v_0}{c}$$

where λ_0 is the emitted wavelength, v_0 is the velocity of the object and λ is the observed wavelength. Note that $v<0$ corresponds to $\lambda<\lambda_0$, a blueshifted wavelength. A redshifted absorption or emission line will appear more towards the red end of the spectrum than a stationary line. In 1913 Vesto Slipher determined the Andromeda Galaxy was blueshifted, meaning it was moving towards the Milky Way. He recorded the spectra of 20 other galaxies — all but 4 of which were redshifted — and was able to calculate their velocities relative to the Earth. Edwin Hubble would later use this information, as well as his own observations, to define Hubble's law: The further a galaxy is from the Earth, the faster it is moving away from us. Hubble's law can be generalised to:

$$v = H_0 d$$

where v is the velocity (or Hubble Flow), H_0 is the Hubble Constant, and d is the distance from Earth.

Redshift (z) can be expressed by the following equations:

Calculation of redshift, z	
Based on wavelength	Based on frequency
$z = \frac{\lambda_{obsv} - \lambda_{emit}}{\lambda_{emit}}$	$z = \frac{f_{emit} - f_{obsv}}{f_{obsv}}$
$1 + z = \frac{\lambda_{obsv}}{\lambda_{emit}}$	$1 + z = \frac{f_{emit}}{f_{obsv}}$

In these equations, frequency is denoted by f and wavelength by λ. The larger the value of z, the more redshifted the light and the farther away the object is from the Earth. As of January 2013, the largest galaxy redshift of z~12 was found using the

Hubble Ultra-Deep Field, corresponding to an age of over 13 billion years (the universe is approximately 13.82 billion years old).

The Doppler effect and Hubble's law can be combined to form the equation $z = \frac{v_{\text{Hubble}}}{c}$, where c is the speed of light.

Peculiar Motion

Objects that are gravitationally bound will rotate around a common center of mass. For stellar bodies, this motion is known as peculiar velocity, and can alter the Hubble Flow. Thus, an extra term for the peculiar motion needs to be added to Hubble's law:

$$v_{\text{total}} = H_0 d + v_{\text{pec}}$$

This motion can cause confusion when looking at a solar or galactic spectrum, because the expected redshift based on the simple Hubble law will be obscured by the peculiar motion. For example, the shape and size of the Virgo Cluster has been a matter of great scientific scrutiny due to the very large peculiar velocities of the galaxies in the cluster.

Binary Stars

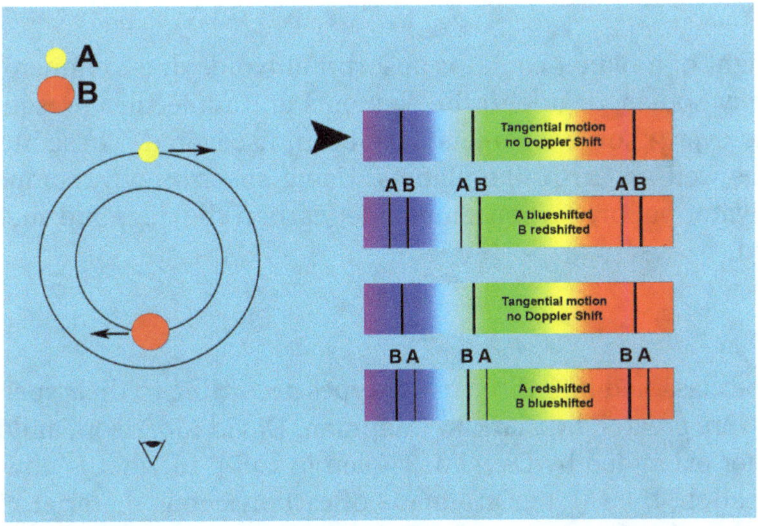

Two stars of different size orbiting the center of mass. The spectrum can be seen to split depending on the position and velocity of the stars.

Just as planets can be gravitationally bound to stars, pairs of stars can orbit each other. Some binary stars are visual binaries, meaning they can be observed orbiting each other through a telescope. Some binary stars, however, are too close together to be resolved. These two stars, when viewed through a spectrometer, will show a composite spectrum: the spectrum of each star will be added together. This composite spectrum becomes easier to detect when the stars are of similar luminosity and of different spectral class.

Spectroscopic binaries can be also detected due to their radial velocity; as they orbit around each other one star may be moving towards the Earth whilst the other moves away, causing a Doppler shift in the composite spectrum. The orbital plane of the system determines the magnitude of the observed shift: if the observer is looking perpendicular to the orbital plane there will be no observed radial velocity. For example, if you look at a carousel from the side, you will see the animals moving toward and away from you, whereas if you look from directly above they will only be moving in the horizontal plane.

Planets, Asteroids and Comets

Planets, asteroids, and comets all reflect light from their parent stars and emit their own light. For cooler objects, including solar-system planets and asteroids, most of the emission is at infrared wavelengths we cannot see, but that are routinely measured with spectrometers. For objects surrounded by gas, such as comets and planets with atmospheres, further emission and absorption happens at specific wavelengths in the gas, imprinting the spectrum of the gas on that of the solid object. In the case of worlds with thick atmospheres or complete cloud cover (such as the gas giants, Venus, and Saturn's satellite Titan (moon)), the spectrum is mostly or completely due to the atmosphere alone.

Planets

The reflected light of a planet contains absorption bands due to minerals in the rocks present for rocky bodies, or due to the elements and molecules present in the atmosphere. To date over 3,500 exoplanets have been discovered. These include so-called Hot Jupiters, as well as Earth-like planets. Using spectroscopy, compounds such as alkali metals, water vapor, carbon monoxide, carbon dioxide, and methane have all been discovered.

Asteroids

Asteroids can be classified into three major types according to their spectra. The original categories were created by Clark R. Chapman, David Morrison, and Ben Zellner in 1975, and further expanded by David J. Tholen in 1984. In what is now known as the Tholen classification, the C-types are made of carbonaceous material, S-types consist mainly of silicates, and X-types are 'metallic'. There are other classifications for unusual asteroids. C- and S-type asteroids are the most common asteroids. In 2002 the Tholen classification was further "evolved" into the SMASS classification, expanding the number of categories from 14 to 26 to account for more precise spectroscopic analysis of the asteroids.

Comets

The spectra of comets consist of a reflected solar spectrum from the dusty clouds surrounding the comet, as well as emission lines from gaseous atoms and molecules excited to fluorescence by sunlight and/or chemical reactions. For example, the chemical composition of Comet ISON was determined by spectroscopy due to the prominent emission lines of cyanogen (CN), as well as two- and three-carbon atoms (C_2 and C_3). Nearby comets can even be seen in X-ray as solar wind ions flying to the coma are neutralized. The cometary X-ray spectra therefore reflect the state of the solar wind rather than that of the comet.

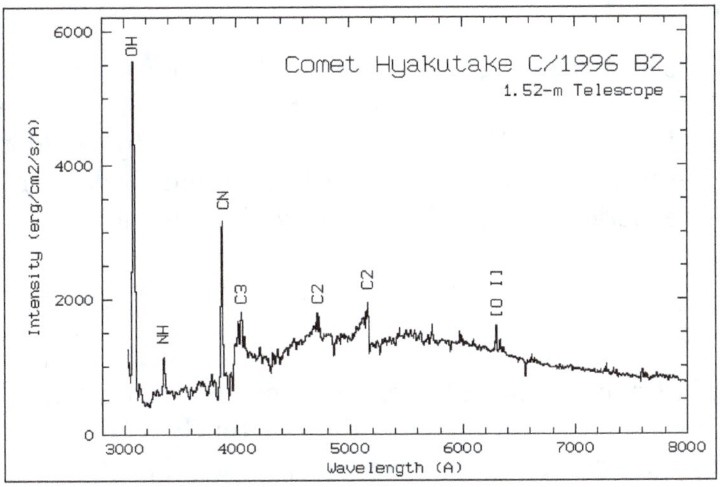

Optical spectrum of Comet Hyakutake.

ASTRONOMICAL CLOCK

An astronomical clock, horologium, or orloj is a clock with special mechanisms and dials to display astronomical information, such as the relative positions of the sun, moon, zodiacal constellations, and sometimes major planets.

The astrarium made by the Italian astronomer and physician Giovanni Dondi dell'Orologio showed hour, year calendar, movement of the planets, Sun and Moon.

The term is loosely used to refer to any clock that shows, in addition to the time of day, astronomical information. This could include the location of the sun and moon in the sky, the age and Lunar phases, the position of the sun on the ecliptic and the current zodiac sign, the sidereal time, and other astronomical data such as the moon's nodes (for indicating eclipses) or a rotating star map. The term should not be confused with astronomical regulator, a high precision but otherwise ordinary pendulum clock used in observatories.

Astronomical clocks usually represent the solar system using the geocentric model. The center of the dial is often marked with a disc or sphere representing the earth, located at the center of the solar system. The sun is often represented by a golden sphere, shown rotating around the earth once a day around a 24-hour analog dial. This view accorded both with the daily experience and with the philosophical world view of pre-Copernican Europe.

Le Gros Horloge

Le Gros Horloge in Rouen is one of the earliest known astronomical clocks. The clock is installed in a Renaissance arch crossing the Rue du Gros-Horloge. The mechanism is one of the oldest in France; the movement was made in 1389. Construction of the clock was started by Jourdain del Leche, who lacked the necessary expertise to finish the task, so the work was completed by Jean de Felain, who became the first to hold the position of governor of the clock. The clock was originally constructed without a dial, with one revolution of the hour-hand representing twenty-four hours. The movement is cast in wrought iron, and at approximately twice the size of the Wells Cathedral clock, it is perhaps the largest such mechanism still extant. A facade was added in 1529 when the clock was moved to its current position. The Renaissance facade represents a golden sun with 24 rays on a starry blue background. The dial measures 2.5 metres in diameter.

Le Gros Horloge.

The phases of the moon are shown in the oculus of the upper part of the dial. It completes a full rotation in 29 days. The week days are shown in an opening at the base of the dial with allegorical subjects for each day of the week.

Su Sung's Cosmic Engine

The Science Museum (London) has a scale model of the 'Cosmic Engine', which Su Sung, a Chinese polymath, designed and constructed in China in 1092. This great astronomical hydromechanical clock tower was about ten metres high (about 30 feet) and featured a clock escapement and was indirectly powered by rotating wheel either with falling water and liquid mercury, which freezes at a much lower temperature than water, allowing operation of the clock during colder weather. A full-sized working replica of Su Sung's clock exists in the Republic of China (Taiwan)'s National Museum of Natural Science, Taichung city. This full-scale, fully functional replica, approximately 12 meters (39 feet) in height, was constructed from Su Sung's original descriptions and mechanical drawings.

Al-Jazari's Castle Clock

The most sophisticated water-powered astronomical clock was Al-Jazari's castle clock, considered to be an early example of a programmable analog computer, in 1206. It was a complex device that was about 33 meters (108 feet) high, and had multiple functions alongside timekeeping. It included a display of the zodiac and the solar and lunar orbits, and a pointer in the shape of the crescent moon which travelled across the top of a gateway, moved by a hidden cart and causing automatic doors to open, each revealing a mannequin, every hour.

It was possible to re-program the length of day and night every day in order to account for the changing lengths of day and night throughout the year, and it also featured five musician automata who automatically played music when moved by levers operated by a hidden camshaft attached to a water wheel. Other components of the castle clock included a main reservoir with a float, a float chamber and flow regulator, plate and valve trough, two pulleys, crescent disc displaying the zodiac, and two falcon automata dropping balls into vases.

Strasbourg

The Strasbourg Cathedral has housed three different astronomical clocks since the 14th century. The first clock was built between 1352 and 1354 and stopped working sometime at the beginning of the 16th century. A second clock was then built by Herlin, Conrad Dasypodius, the Habrecht brothers, and others, between 1547 and 1574. This clock stopped working in 1788 or 1789 (as it apparently stopped working gradually, each component being disconnected one after the other). After a lapse of 50 years, a new clock was built by Jean-Baptiste Schwilgué and about 30 workers. This clock is housed in the case of the 2nd clock. It shows many astronomical and calendrical functions (including what is thought to be the first complete mechanization of the part of the computus needed to compute Easter) as well as several automata.

Strasbourg astronomical clock.

Prague Astronomical Clock.

Prague

One of the most famous of this type of clock is the Old-Town Hall clock in Prague, Czech Republic. It is also known as the Prague orloj. The central portion was completed in 1410. The four figures are set in motion at the hour, with Death (represented by a skeleton) striking the time. On the hour there is a presentation of statues of the Apostles at the doorways above the clock, with all twelve presented at noon. In 1490 a calendar display was added below the clock along with decorative Gothic sculptures.

During World War II the clock was nearly destroyed by Nazi fire. The townspeople are credited with heroic efforts in saving most of the parts. It was gradually renovated until 1948. In 1979 the clock was once more cleaned and renovated. According to local legend the city will suffer if the clock is neglected and its good operation is placed in jeopardy.

Olomouc

Olomouc, the former capital of Moravia in the eastern part of the Czech Republic, also has an impressive exterior astronomical clock on the main town square. It is a rare example of a heliocentric astronomical clock.

A legend dates its construction to the year 1422; however, in historic sources it is first mentioned in 1517. The clock was remodelled approximately once every century; in 1898 the astrolabe was replaced with a heliocentric model of the solar system. When the retreating Nazi German army passed through Olomouc in the final days of the war in May 1945, they opened fire on the old astronomical clock, leaving only a few pieces (that can now be seen in the local museum). As a result of the serious damage the clock was reconstructed in the style of socialist-realism in the first years of communist rule in Czechoslovakia. The religious and royal figures were replaced with athletes, workers, farmers, scientists and other members of the proletariat, while the glockenspiel was altered to play three pieces of traditional local music.

Olomouc astronomical clock detail.

Stará Bystrica

Stará Bystrica astronomical clock.

In 2009, an astronomical clock was built in the municipality of Stará Bystrica in Northern Slovakia during the reconstruction of the town square.

The astronomical clock has the shape of a stylized form of Our Lady of Sorrows, patron of Slovakia; it has been described as the largest wooden statue of Slovakia. Its exterior is decorated by statues of important figures from Slovakia's history: Prince Pribina, King Svatopluk, Anton Bernolák, Ľudovít Štúr, Milan Rastislav Štefánik, Andrej Hlinka. Each hour, statuettes of saints connected with Slovakia appear: Cyril, Methodius, Andrew-Zorard, Benedict, Gorazd, Bystrík and Adalbert. The bells of the clock carry the names Sv. Juraj (St. George) and Riečnická Madona (Our Lady of Riečnica); the first is rung to indicate the time, the second accompanies the saints. This astronomical clock is the only one in Slovakia.

The astronomical part of the clock consists of an astrolabe displaying the astrological signs, positions of the Sun and Moon, and the lunar phases. The clock is controlled by computer using DCF77 signals.

References

- Nahar, Anil K. Pradhan, Sultana N. (2010). Atomic astrophysics and spectroscopy. Cambridge: Cambridge University Press. pp. 7, 221. ISBN 978-0-521-82536-8
- Astronomicaltelescope, opticalinstruments, Physics, study-zone: funscience.in, Retrieved 21 April, 2020
- Ball, David W. (2001). Basics of Spectroscopy. Bellingham, Washington: Society of Photo-Optical Instrumentation Engineers. pp. 24, 28. ISBN 0-8194-4104-X
- Radio-telescopes, chapter, suny-astronomy: lumenlearning.com, Retrieved 17 February, 2020
- Kuchner and Traub (2002). "A Coronagraph with a Band-limited Mask for Finding Terrestrial Planets". The Astrophysical Journal. 570 (2): 900–908. Arxiv:astro-ph/0203455. Bibcode:2002apj...570..900K. Doi:10.1086/339625
- Visible-light-detectors-and-instruments, chapter, suny-astronomy: lumenlearning.com, Retrieved 24 March, 2020

PERMISSIONS

All chapters in this book are published with permission under the Creative Commons Attribution Share Alike License or equivalent. Every chapter published in this book has been scrutinized by our experts. Their significance has been extensively debated. The topics covered herein carry significant information for a comprehensive understanding. They may even be implemented as practical applications or may be referred to as a beginning point for further studies.

We would like to thank the editorial team for lending their expertise to make the book truly unique. They have played a crucial role in the development of this book. Without their invaluable contributions this book wouldn't have been possible. They have made vital efforts to compile up to date information on the varied aspects of this subject to make this book a valuable addition to the collection of many professionals and students.

This book was conceptualized with the vision of imparting up-to-date and integrated information in this field. To ensure the same, a matchless editorial board was set up. Every individual on the board went through rigorous rounds of assessment to prove their worth. After which they invested a large part of their time researching and compiling the most relevant data for our readers.

The editorial board has been involved in producing this book since its inception. They have spent rigorous hours researching and exploring the diverse topics which have resulted in the successful publishing of this book. They have passed on their knowledge of decades through this book. To expedite this challenging task, the publisher supported the team at every step. A small team of assistant editors was also appointed to further simplify the editing procedure and attain best results for the readers.

Apart from the editorial board, the designing team has also invested a significant amount of their time in understanding the subject and creating the most relevant covers. They scrutinized every image to scout for the most suitable representation of the subject and create an appropriate cover for the book.

The publishing team has been an ardent support to the editorial, designing and production team. Their endless efforts to recruit the best for this project, has resulted in the accomplishment of this book. They are a veteran in the field of academics and their pool of knowledge is as vast as their experience in printing. Their expertise and guidance has proved useful at every step. Their uncompromising quality standards have made this book an exceptional effort. Their encouragement from time to time has been an inspiration for everyone.

The publisher and the editorial board hope that this book will prove to be a valuable piece of knowledge for students, practitioners and scholars across the globe.

INDEX

www.ingramcontent.com/pod-product-compliance
Lightning Source LLC
LaVergne TN
LVHW080626210926
848485LV00003B/216